Praise for *Saving Us*

"A welcome effort to address the politicization of climate change head-on . . . Hayhoe is an appealing messenger."

—*Financial Times*

"Practical advice abounds in this compassionate guide to conducting meaningful discussions about the environment. Those in search of a hope-filled approach will find plenty of encouragement."

—*Publishers Weekly*

"This heartfelt and empathetic book focuses on shared values . . . to build essential bonds and the courage required for the work ahead. Read this vital book and take its exigent message to heart."

—*Washington Independent Review of Books*

"A must-read if we're serious about enacting positive change from the ground up, in communities, and through human connections and human emotions."

—Margaret Atwood

"Bold and pragmatic, *Saving Us* is a vital contribution to the discussion on climate change."

—Chelsea Clinton, *New York Times* bestselling author and global health advocate

"Conversations fueled by respect and shared values can help save our planet, and Katharine Hayhoe gives us the confidence to do what it takes."

—Alan Alda, Emmy Award–winning actor and host of *Clear+Vivid with Alan Alda*

"Katharine shares an optimistic outlook on what we all can do to move the needle toward solutions and invite allies under the big tent."
—Don Cheadle, Academy Award–nominated actor and UN Environmental Program Goodwill Ambassador

"*Saving Us* contains profound insights on human behavior, and it shows us how our conversations can launch us on the journey away from despair toward awareness and engagement. A real joy to read."
—Christiana Figueres, executive secretary of the UN Framework Convention on Climate Change

"*Saving Us* offers a road map to transform our approach to tackling climate challenges from sprawling global crises to community-driven solutions, recognizing that our diverse and collective voices are key to creating lasting change."
—Abby Maxman, president and CEO of Oxfam America

"Dr. Hayhoe writes personally and persuasively—as a person of faith and as a scientist—about both the peril of the climate crisis and why we can still have hope."
—Dan Misleh, executive director of the Catholic Climate Covenant

"A masterful playbook exploring why past approaches have failed, and how we can all help get it right.
—Professor Dave Reay, chair in Carbon Management and executive director of Edinburgh Climate Change Institute

SAVING US

A CLIMATE SCIENTIST'S CASE FOR *HOPE* AND *HEALING* IN A DIVIDED WORLD

KATHARINE HAYHOE

ONE SIGNAL
PUBLISHERS

———

ATRIA

NEW YORK   LONDON   TORONTO   SYDNEY   NEW DELHI

ONE SIGNAL
PUBLISHERS

ATRIA

An Imprint of Simon & Schuster, Inc.
1230 Avenue of the Americas
New York, NY 10020

Some names have been changed.

First One Signal Publishers/Atria Paperback edition September 2022

ONE SIGNAL PUBLISHERS / ATRIA PAPERBACK and colophon are trademarks of Simon & Schuster, Inc.

For information about special discounts for bulk purchases, please contact Simon & Schuster Special Sales at 1-866-506-1949 or business@simonandschuster.com.

The Simon & Schuster Speakers Bureau can bring authors to your live event. For more information or to book an event, contact the Simon & Schuster Speakers Bureau at 1-866-248-3049 or visit our website at www.simonspeakers.com.

Interior design by Jill Putorti

Manufactured in the United States of America

1 3 5 7 9 10 8 6 4 2

Library of Congress Control Number: 2021940932

ISBN 978-1-9821-4383-1
ISBN 978-1-9821-4384-8 (pbk)
ISBN 978-1-9821-4385-5 (ebook)

*To everyone who believes the difficult
issues in life are worth talking about*

CONTENTS

SECTION 5:
YOU CAN MAKE A DIFFERENCE

PREFACE

It took a pandemic to bring us together.

As coronavirus swept across the world, we saw country after country go into lockdown. Schools closed and places of work shut down. For a while, it seemed we were all united against a common threat.

As the pandemic wore on, this consensus started to disintegrate.

Political leaders who'd been elected on waves of nationalist and populist rhetoric began to depict coronavirus precautions as more damaging to society than the disease itself. Supposedly reputable sources hyped false cures, and deliberately misrepresented the severity of the illness. In the U.S., going without a face mask became a badge of the conservative cause. While some continued to social distance, others held gatherings, some of which turned into superspreader events. The contrast between nations was stark—and the difference in infection and death rates equally so.

Sadly, none of this surprised me. I've spent my career studying climate change. The same techniques used to politicize coronavirus—promoting pseudoscience and fake experts, slandering the actual experts, valuing the economy over human life, even hiding or denying data to make the issue seem less urgent and less harmful than it is—have been applied to climate change for decades. Horrified, I've seen political biases drive people to reject simple facts: climate is changing, humans are responsible, the impacts are serious, and the time to act is now.

The U.S. is arguably home to the most extreme divisions between liberal and conservative, but I've seen these same divisions growing in recent years in my home country of Canada, in the U.K., in Australia, in Europe, and beyond. No matter where we live, the result is the same: as people identify with increasingly narrow tribes, they begin to view those with different views as alien, not worth respecting or even treating as human. The Beyond Conflict Institute's 2020 report, *America's Divided Mind*, didn't mince any words: "Increasingly, Americans who identify themselves as either Democrats or Republicans view one another less as fellow citizens and more as enemies who represent a profound threat to their identities."

During the last full year of Donald Trump's Republican administration, the U.S. experienced the world's highest death toll from the pandemic. His time in office also coincided with apocalyptic wildfires, record-breaking hurricanes, and dire scientific warnings of climate impacts. Yet in the November 2020 election, Trump's base stuck with him. Democrat Joe Biden eked out a narrow win in key swing states. The country remained as divided as ever.

Religion, politics, and money have long been potentially combustible topics. Today, though, climate change tops that list. It's the most politicized and divisive issue in the U.S. At this point, overcoming the polarization seems nearly hopeless, and solving climate change, even more so. Is there any remedy to our deep-rooted and growing divide?

———

If you're already worried about climate change and support climate action, you're not alone. Over 50 percent of adults in the U.S. are alarmed or concerned about climate change. People in other countries are even more concerned—from 67 percent in Australia to over 90 percent in countries like Thailand, the Philippines, and Vietnam. Around the world, "there is a strong sense [that all] have the power to combat climate change." In the U.S., seven out of ten say they wish they could do something to fix it; but half of them don't know where to start and only 35 percent say they ever talk about it, even occasionally.

I'm a scientist who studies climate change. But increasingly, I spend more and more of my time explaining *why* it matters: to Christians at church groups and kids at science museums, businesspeople at conferences and community members at neighborhood meetings.

Why? Because after thousands of conversations, I'm convinced that the single most important thing that anyone—not just me, but literally anyone—can do to bring people together is, ironically, the very thing we fear most. *Talk about it.*

Why are people not talking about something that matters to them so much? Even if we agree it's real and it's serious, talking about it can be discouraging and depressing. There's too great a risk the conversation might devolve into a screaming match or end up leaving everyone overwhelmed by the enormity of the problem. We want to talk about it; we just don't know how.

"*How* do I talk about this . . . to my mother, brother-in-law, friend, colleague, neighbor, elected official?" I'm asked this question nearly everywhere I go.

A fellow scientist asks me this after a lecture in a wood-paneled theater at UC-Berkeley in California. In Slovakia, I listen in as a group debates it in the upper balcony of the old, stripped-down city hall at the country's first national climate meeting. Moms, dads, and teachers post on my Facebook page wanting to know what to say to their kids. My friend texts me asking how to reply to the disinformation his mom just shared with him.

Usually, they've already given conversation a try. They've boned up on a few alarming scientific facts. They've tried to explain how fast the Arctic is melting, or how bees are disappearing, or how carbon dioxide levels are rising. But their attempts have fallen flat. Why? Because the biggest challenge we face isn't science denial. It's a combination of tribalism, complacency, and fear. Most don't think climate change is going to affect them personally or that we can do anything reasonable to fix it; and why would they, if we never talk about it?

It's important to understand what's happening to our world and how it affects us. But bombarding someone with more data, facts, and science only engages their defenses, pushes them into self-justification, and leaves us more divided than when we began. On climate change and other issues with moral implications, we tend to believe that everyone should care for the same self-evident reasons we do. If they don't, we all too often assume they lack morals. But most people do have morals and are acting according to them; they're just different from ours. And if we are aware of these differences, we can speak to them.

Here's more good news. The Beyond Conflict Institute's report also shows that people perceive that "the other side" disagrees with them far more than is actually the case. So instead of reacting to something you disagree with, what if you started a conversation about something you agree on? What if you asked questions rather than arguing? What if you shared, genuinely and personally, how climate change threatens what you care about? And what if you talked about practical, real-world solutions that are already available today?

Beginning a conversation with something that unites us instead of something that divides us means we are starting at a place of mutual respect, agreement, and understanding—which is pretty much the opposite of where most conversations about contentious issues like climate change begin these days. And as we truly listen, we're likely to discover more surprising points of agreement.

In this book I want to show you how to have conversations that will help you to reconnect with family and friends in real life, building genuine relationships and communities rather than tribes and bubbles. The bald facts are scary, and necessary. But climate change connects to the things we all care about: the health of our families, the economic strength of our communities, and the stability of our world. Fixing it isn't only good for the planet; it's good for all of us, too.

The bottom line is this. To care about climate change, you only need to be one thing, and that's a person living on planet Earth who wants a better future. Chances are, you're already that person—and so is everyone else you know.

THE PROBLEM AND THE SOLUTION

1

DEMOCRATS AND DISMISSIVES

"It is a common folk theory . . . that facts will set you free."

GEORGE LAKOFF, DON'T THINK OF AN ELEPHANT!

"Climate change is the second biggest hoax after the corona scamdemic."

MAN ON TWITTER TO KATHARINE

I'm getting used to being hated. It's not for anything I've done; it's because of what I represent. Communist, libtard, lunatic; Jezebel, liar, and whore; high priestess of the climate cult and handmaiden of the Antichrist, I've been called it all.

We scientists can take criticism and give it, too. Our professional exchanges and reviews of one another's work don't pull any punches. Yet it's hard not to find such epithets disturbing. Even more unsettling, they seem to come out of nowhere, and offer no clear avenue of response. If a colleague disagrees with my ideas, it motivates me to collect more and better data—data that sometimes shows they're right. That's how science works. But what am I supposed to do when I'm called a "climate ho"—somehow prove I'm not?

Much of it arrives virtually, but the first time I faced this attitude was in person. During my first year as an atmospheric science professor in Texas, a colleague asked me to guest-teach his early morning undergraduate geology class. It was a challenging time of day to ask anyone to absorb the details of how carbon moves through the planet's climate system. Still, I optimistically connected my laptop to the projector and peered out into the dark, cavernous lecture hall. Most of the seats were full, so I launched into my carefully prepared presentation.

Every teacher thinks their favorite topic is fascinating, and I was no exception. How could you not want to understand the history of our world? But my talk just didn't seem to capture the students' interest. A few took notes, but most seemed to be checking Facebook on their laptops or sneaking a nap. Even the last few minutes of the lecture, where I described how humans had accelerated the natural carbon cycle by millions of years through digging up and burning fossil fuels, didn't get any reaction.

Trying to hide my disappointment, I called for questions. A tall, athletic-looking student raised his hand and stood up so I could see him. I nodded eagerly. Then in a belligerent tone, he demanded,

"You're a Democrat, aren't you?"

Floored, I replied, "No—I'm Canadian!"

That was a relatively benign introduction to what's now become a regular part of my life. Nearly every day I receive angry, even hate-filled, objections to the work I do as a climate scientist: tweets, Facebook comments, even the occasional phone call or handwritten letter. "You make your living off climate hysteria," reads one tweet. A multipage, single-spaced manifesto in my university mailbox starts, "You Lie!!!!" A Facebook message screams, "Get aborted you human-hating c***." But before I block anyone on social media, I look at their profile. I want to know what type of person would go out of their way to write things like this to someone they don't even know.

About a third of the social media accounts that hurl insults score high on the online bot ratings, indicating they probably aren't real people. They're just part of the automated online attack squad that's regularly aimed at everyone from progressive politicians to COVID virologists. But most other accounts seem to be associated with living, breathing humans. If they are from the U.S., their profile nearly always features the acronyms MAGA, KAG, or QAnon, all trademarks of right-wing ideology. If they are from Canada, they usually hate Prime Minister Justin Trudeau, love Alberta's oil and gas industry, and support the Conservative or, increasingly, the ultra-right-wing People's Party of Canada. If they're from the U.K., they're pro-Brexit. If they're from Australia, they're likely to support the conservative Prime Minister Scott Morrison. Wherever they're from, they want everyone to know how much they love their country and how much they hate the proponents of political correctness, the mainstream media, and the "leftards and commies" who are intent on destroying it.

HOW DID CLIMATE CHANGE BECOME SO POLARIZED?

A thermometer doesn't give you a different answer depending on how you vote. Even in the U.S., climate change used to be a respectably bipartisan issue well within most of our lifetimes. In 1998, a Gallup poll found that 47 percent of Republicans and 46 percent of Democrats agreed that the effects of global warming had already begun. In 2003, Senators John McCain, an Arizona Republican, and Joseph Lieberman, then a Connecticut Democrat, introduced the Climate Stewardship Acts. As recently as 2008, former speaker of the House Newt Gingrich, a Republican, and current House speaker Nancy Pelosi, a Democrat, cozied up on a love seat in front of the U.S. Capitol to film a commercial about climate change. "We do agree, our country must take action to address climate change," Gingrich said, while Pelosi added, "We need cleaner forms of energy, and we need them fast." But for the past decade, climate change has topped the list of America's most polarized issues, along with immigration, gun policy, and race relations. By 2020, coronavirus had joined the list.

It's not just the U.S. In Canada, there's nearly a one-to-one correspondence between people's response to the question "Is the Earth warming?" and the party that won that particular area in the 2019 federal election. The more conservative Canadian voters are, the more likely they are to reject what a hundred and fifty years of temperature data is telling us. In the U.K., Conservative members of parliament are five times more likely to vote against climate legislation than their Labour counterparts. In Australia, the influence of the coal industry and the Murdoch-controlled press on national politics is undeniable. Australia was the first and only country to implement, then withdraw, a carbon tax after just two years. More recently, some of its politicians asserted that climate change had nothing to do with the devastating wildfires in late 2019 and early 2020. Their claims were bolstered by disinformation—including that the fires were started by climate activists—that was deliberately introduced and circulated on social media in a manner similar to "past disinformation campaigns, such as the coordinated behavior of Russian trolls during the 2016 U.S. presidential election," one study found.

Although science denial dominates the headlines, people's rejection of the science on climate change is rarely about the science itself. In a study of fifty-six countries, researchers found people's opinions on climate change to be most strongly correlated not with education or knowledge, but rather with "values, ideologies, worldviews and political orientation." Like coronavirus, vaccines, and more, climate denial is often just one part of a toxic stew of identity issues that share a key factor: fear of change. Societal change is happening faster today than at any time in our lifetimes, and many are afraid they're already being left behind. That fear drives tribalism, emphasizing what divides us rather than what unites us; and the more threatened we feel, the tighter we draw the circles to distinguish between *them* and *us*.

That's why so much of the polarization is emotional. Over the past forty years, researchers Greg Lukianoff and Jonathan Haidt found that "people [in the U.S.] who identify with either of the two main political parties increasingly hate and fear the other party and the people in it." This polarization can also be literally mind-altering. In one experiment, when asked what they thought of an issue, then told that their political party thought something different about it, many immediately changed their opinions *and were unaware of the fact that they were doing so.*

A lot of what we see online doesn't help. "As media frames opposing viewpoints as shouting matches and comments on Facebook and Twitter convey vitriol and accusation," says psychologist Tania Israel, "we shy away from people and organizations whose positions may conflict with our own. We take refuge in echo chambers of like-minded people expressing views we support, cheering each other on as we rake our common enemies over the coals."

Giving up intellectual adherence to a scientific issue like climate change that seems remote and virtually unfixable seems like a small price to pay to be part of a tribe that accepts us and makes us feel safe. It may even be a benefit, because who really wants to believe that climate change could spell the end of our civilization? And the more we experience the benefits of belonging, the more willing we are to tailor our beliefs to those of our tribe.

WHY TWO CLIMATE TRIBES ARE NOT ENOUGH

We often assume that the tribes that form around climate change can be sorted into two categories: *them* and *us*. In reality, though, it's a lot more complicated than that. I also have a problem with the labels that are most often applied to those categories: *believers* and *deniers*.

I object to "believers" because climate change is not, at its core, a matter of faith. I don't "believe" in science: I make up my mind based on facts and data, much of which can be seen and shared. Not only that, but climate change is often deliberately—and very successfully—framed as an alternate, Earth-worshipping religion. This is sometimes subtle, as the church sign that reads, "On Judgement Day, you'll meet Father God not Mother Earth." Other times this point is made much more explicitly, like when Senator Ted Cruz told Glenn Beck in 2015 that "climate change is not science, it's religion," and Senator Lindsey Graham said in 2014 that "the problem is Al Gore's turned this thing into religion."

And while it may be convenient for some climate advocates to dismiss their opponents as "deniers," it's an unhelpful label if you want to win people over. I've also seen it applied all too often to shut down discussion, rather than encourage it, through stereotyping and dismissing anyone who expresses any doubts about the reality of climate change.

Instead, I prefer the classification system created by researchers Tony Leiserowitz and Ed Maibach. Called *Global Warming's Six Americas*, it divides people into six groups rather than just two. Tony and Ed have tracked changes in these groups nationally since 2008. At one end of the spectrum, there are the Alarmed, the only group that has grown significantly since they began the study. The Alarmed are convinced global warming is a serious and immediate threat but many still don't know what to do about it. In 2008, they made up just 18 percent of the U.S. population. By the end of 2019, they had reached 31 percent, before falling back to 26 percent in 2020. The next group, the Concerned, also accept the science and support climate policies, but see the threat as more distant. They started at 33 percent in 2008 and moved down to 28 percent by 2020 as more became Alarmed. The number of the Cautious, who still need to be convinced that the problem is real, serious, and urgent, has remained steady around 20 percent. The Disengaged are people who know little and care less. They've gone from 12 percent in 2008 to 7 percent in 2020. Next there are the 11 percent of Americans who remain Doubtful and don't consider climate change a serious risk, or consider it much at all. Finally, at the far end of the spectrum, there are the 7 percent who remain Dismissive. Angrily rejecting the idea that human-caused climate change is a threat, they are most receptive to misinformation and conspiracy theories.

INTRODUCING THE DISMISSIVE

You might know a Dismissive. A Dismissive is someone who will discount any and every thing that might show climate change is real, humans are responsible, the impacts are serious, and we need to act now. In pursuit of that goal, they will dismiss hundreds of scientific experts, thousands of peer-reviewed scientific studies, tens of thousands of pages of scientific reports, and two hundred years of science itself.

Dismissives can't leave the topic of climate change alone. They're constantly commenting on Facebook posts, talking about it at family dinners, forwarding articles they've found that buttress their point. They may go out of their way to ridicule people who support climate action and environmentally friendly behavior, such as driving fuel-efficient cars, installing solar panels, and adopting plant-based diets. They quote blogs that peddle pseudoscience claiming that the Antarctic ice sheet is growing, or that scientists are faking global temperature data. Dismissives dominate the comment section of online articles and the op-ed pages of the local newspaper. They account for most of the attacks I receive on social media, too.

Because of their obsession with the topic, when we dream about having a constructive conversation with someone about climate change, often a Dismissive is the first person who comes to mind. Unfortunately, though, the "seven-percenters," as I think of them, are the only ones it's nearly impossible to have a positive conversation with. Here's why.

My uncle is a Dismissive. For a long time, he voiced his objections to climate change at family reunions and in conversations with my dad. But last year, he decided to email them directly to me. His arguments were nothing new: they challenged the basic physics of heat-trapping gases and blamed climate change on natural factors, not humans. So I replied to his email with detailed sources explaining the physics scientists have understood since the 1800s. I also sent him some articles from the helpful Skeptical Science website debunking the arguments he'd raised.

I imagined it would take my uncle at least a few days to wade through and consider the resources I'd provided. Instead, he responded almost immediately, dismissing what I'd sent and voicing even more arguments. What had happened? I had fallen right into the trap of believing that facts could convince someone whose identity is built on rejecting climate science.

We often believe that "if we just tell people the facts, since people are basically rational beings, they'll all reach the right conclusions," cognitive linguist George Lakoff explains. But that's not the way we humans think. Instead, we think in what he calls "frames." Frames are cognitive structures that determine how we see the world. When we encounter facts that don't fit our frame, it's the frame that stays while "the facts are either ignored, dismissed, [or] ridiculed."

For a Dismissive, disagreeing with the science of climate change is one of their strongest frames. It's so integral to who they are that it renders them literally incapable of considering something they think would threaten their identity. Time and time again on social media I've seen Dismissives refuse point-blank to even click a link that answers the question they've posed to me. And while I believe I've witnessed a few miraculous conversions, so to speak, I don't believe my arguments had much to do with making them happen.

So I didn't send my uncle any more resources. And now when people ask me how I—or they—can convince a Dismissive, their parent or their colleague or their in-law or their elected official, my answer is typically "You probably can't. But the good news is that 93 percent of us aren't Dismissive. They're the ones we *can* have constructive conversations with. They're the ones who can make a difference."

HOW TO BREAK THE CLIMATE CYCLE

How do we talk about climate change constructively with the 93 percent? Unfortunately, our instincts can lead us astray here, too. As we get more and more worried, we often feel compelled to dump scary data on people so they will share our fear. Scientists publish report after report warning of melting ice sheets, scorching heat waves, devastating rainfall events, unprecedented wildfires, and ever more powerful hurricanes. We desperately want to sound the alarm—and we're not wrong. Climate change *is* alarming. But our natural reaction often makes the situation worse, not better.

Research on everything from airplane seatbelts to hand washing in hospitals shows that bad-news warnings are more likely to make people check out than change their behavior. And the more vivid and dire the picture painted, the less responsive the recipient. "Fear and anxiety [can] cause us to withdraw, to freeze, to give up, rather than take action," neuroscientist Tali Sharot explains in her book *The Influential Mind*.

So if arguing with the 7 percent who are Dismissives and dumping more scary information on the other 93 percent of people doesn't really work, is there anything that does?

Yes, there is.

Start with something you have in common. Connect it to why climate change matters to us personally—not the human race in its entirety or the Earth itself, but rather us as individuals. Climate change affects nearly everything that we already care about. It will make us and our children less healthy, our communities less prosperous, and our world less stable. Often, in fact, it already has.

Then, describe what people can and are doing to fix it. There are all kinds of solutions, from cutting our own food waste to powering buses with garbage to using solar energy to transform the lives of some of the poorest people in the world. There are solutions that clean up our air and our water, grow local economies, encourage nature to thrive, and leave us all better off, not worse. Who doesn't want that?

This book is packed with stories, ideas, and information that will lead to positive conversations—conversations that bridge gaps rather than dig trenches, conversations that may surprise you with the discovery of common ground.

By bonding over the values we truly share, and by connecting them to climate, we can inspire one another to act together to fix this problem. But it all begins with understanding who we already are, and what we already care about—because chances are, whatever that is, it's already being affected by climate change, whether we know it or not.

WHO I AM

"Climate change public communication and engagement efforts must start with the fundamental recognition that people are different and have different psychological, cultural, and political reasons for acting."

TONY LEISEROWITZ AND ED MAIBACH, *GLOBAL WARMING'S SIX AMERICAS* **2009**

"Those people who talk about global warming, I don't agree with them at all! But this? This makes sense."

WOMAN AT TEXAS WATER CONSERVATION AUTHORITY MEETING TO KATHARINE

It's nearly impossible to make someone care about climate change for the same reasons I do. But I don't think I have to, and you don't, either. Through thousands of conversations, I've become truly convinced that nearly everyone already *has* the values they need to care about the future of our world, even if they're not the same as mine or yours. And if they don't *think* they care, it's because they just haven't connected the dots. When they do, they'll see for themselves that caring about climate change is entirely consistent with who they are. Climate action, in fact, can be an even more genuine expression of their identity and their values than inaction or denial would be.

As I walked into the drab hotel ballroom to give a guest talk at our local West Texas Rotary Club's luncheon gathering, a massive banner in the entryway caught my eye. At the top, it said: "The Four-Way Test," followed by the club's guiding principles. To assess the value of the things we think, say, or do, they ask:

One—is it the TRUTH?

Two—is it FAIR to all concerned?

Three—will it build GOODWILL and BETTER FRIENDSHIPS?

And four—will it be BENEFICIAL to all concerned?

"That's climate change, and climate action," I thought, amazed. Is climate change the truth? Absolutely. Is it fair? Absolutely not. In fact, that's exactly why I care about it, because it's so profoundly unfair. Would addressing it build goodwill and be beneficial to all concerned? For sure! Inspired, I skipped the buffet lunch and spent the next twenty minutes furiously reorganizing my presentation into the Four-Way Test.

After lunch, I began. At first, there were a lot of folded arms; but as I worked my way through the Four-Way Test, people leaned forward. Heads started to nod. They recognized their values on the screen.

First, climate change is the truth. Scientists have counted over 26,500 independent lines of evidence—including fruit trees blooming earlier, butterflies moving north, and glaciers melting—showing that yes, the planet is warming. It's the truth.

Second, climate change is not fair. It affects our farmers who've done little to cause it, decreasing their crop yields as climate change brings stronger droughts and heat waves. As one local cotton producer recently told me, "I haven't had a good dryland cotton crop since 2007. Fourteen years of unrelenting summer drought!" And what about the poor who benefit from the projects the Rotary Club funds? They've done virtually nothing to contribute to climate change, yet are suffering the worst of its impacts.

What about climate solutions? Rather than being punitive and harmful, many build goodwill and better friendships and, yes, are beneficial as well. Fort Hood, the largest army base in the U.S. by land area, was located right there in Texas. In 2015, they began to transition from fossil fuels to renewable power. Fort Hood now draws 45 percent of its power from solar and wind, saving taxpayers millions. As of 2020, wind and solar energy employed over thirty-seven thousand people across Texas.

By the end of my talk, most heads were nodding along, and most faces were friendly. Many came up afterward to comment positively or ask further questions. Last in line was a local businessman whom I'd met a few times. He'd always been polite, but he'd never brought up climate change with me before. That day, however, he wore an expression I'd never seen before: he appeared bemused.

"I never thought too much of this whole global warming thing," he confessed—which is a polite Texan way of saying he'd thought it was a load of crap—"*but it passed the Four-Way Test.*" In other words, what could he do? He had no choice. Climate change fit right into his value system. Because he was a Rotarian, he already cared about it; he just hadn't realized it until that day. In fact, caring about climate change made him an even better and more genuine Rotarian than he'd been before.

In everyday life, we don't usually get to walk up to a giant banner displaying the rules someone lives by before launching into a conversation. So to figure out where to open a discussion, take inventory of who you are and what you might have in common with others you know and meet. If you don't know what matters to them, ask. Then listen carefully to what they say. Here's my inventory.

WHERE I LIVE

Home is often central to our identity so one of the first things you can ask is, where do you live? I live in Texas, as did everyone else at that Rotary Club meeting. Like them, I want to know that water will come out when I turn on the tap. I also want to know that my house won't be destroyed by floods, and that our city won't turn into a ghost town because we didn't prepare for stronger, longer droughts as climate changes. I'm pretty sure most people in Texas would feel the same way.

When I was invited to address the Texas Water Conservation Authority's annual meeting a few years ago, I decided to try another new approach. I talked, as always, about how temperature and precipitation were changing, and how securing water supply and managing reservoirs would become critically important in a warming world. But I never mentioned the touchy words "climate" and "change" together.

As I collected my computer from the podium afterward, an older woman in a tweed suit came running up to me. She grabbed my hand and pumped it enthusiastically. "I agree with everything you said!" she exclaimed. "Of course we have to prepare!"

Then she continued, disapprovingly, "Those people who talk about global warming, I don't agree with them at all." My mind boggled at this: I'd shown climate projections for higher and lower temperatures in my presentation. If that wasn't global warming, what was?

"But *this*," she concluded with emphasis, speaking about my presentation, "*this* makes sense." By focusing on what we had in common and avoiding the trigger words that would have turned her off, we were able to agree completely that Texas needed smarter water planning to cope with a warmer world.

WHAT I LOVE DOING

What do you enjoy doing? For many of us, exercising and spending time outdoors is crucial to our mental health. Something as simple as walking around the neighborhood has helped to lift my mood during the coronavirus pandemic. Many of my most cherished and evocative memories are of the outdoors: the pine woods of Ontario I'd run through as a child, the windy expansive view from the Colorado mountain I summitted with my dad, the shifting blue-green waters of the Great Lakes where I'd sail with my grandpa.

As a family, we love to ski and snowboard together. But a study I did for *Sports Illustrated* back in 2007 showed that many lower-elevation and more southern ski resorts may be unable to open consistently as climate changes. Already, conditions are getting warmer and often drier, too. I've experienced this firsthand; living in Texas, our favorite place to ski is New Mexico. Just a few years ago, they experienced a winter so hot and dry, some ski areas weren't able to fully open for the season.

I'm sure there are people who enjoy the outdoors who don't think climate change matters to them. I'm even more sure that most of them would care about it, passionately, if they really understood what's at risk.

Many professional and amateur winter athletes recognize that climate change is affecting their winter recreation. They're using their social media platforms and interviews to explain why we need to tackle it before it's too late. Companies like Burton and Patagonia are coming alongside, interspersing information on their websites about how climate change affects snow and ice between videos about untracked powder in Kazakhstan and photos of first summits in the Andes. If we care about the outdoors, then we care about climate change.

WHERE I'M FROM

What is your home country? I'm Canadian, and—like most of us—I love my country. I'm proud of its diversity, its thriving cities, and its natural beauty. But I also see the real threats climate change poses: the encroaching invasive species, thawing permafrost, bigger wildfires, sinking coastlines, and increasing flood risk that have already taken a toll on our True North.

I grew up in a house without air-conditioning, yet just a few decades later it's now a basic necessity in southern Canada. And though Canada has a reputation for accepting refugees from the farthest ends of the Earth, it's estimated that climate change could displace as many as a billion people by mid-century. That's many times more than our entire population.

If climate change continues unchecked, Canada will struggle to ensure the safety and well-being of its own people and economy, let alone make a dent in the coming global refugee crisis. That's why, if you're a Canadian, you care, too: even if you don't know it. And if you're not, I bet I could come up with just as compelling a list of reasons why citizens of your country care about climate change, too.

THOSE I LOVE

Who do you love? Being a mom is a big part of who I am. People can tell you what it's like to become a parent, but until it happens, it's nearly impossible to understand the reaction you experience the first instant you lay eyes on that red-faced, squalling bundle. It literally felt like something in my chest had rearranged itself, and it's never been the same since.

I care about climate change because of my child's future, and that of his cousins, and my friends' kids, too. And how could any caring parent, grandparent, uncle, or aunt not care, if they truly understood the risks?

Organizations like the Mom's Clean Air Force advocate for clean air and a safe world for our kids. I'm also part of Science Moms, where scientists who are moms share with other moms what we know and how we can raise our voices to support climate action and protect our kids' future. By definition we all care, because we love our kids and want this planet to continue to be a safe home for them.

WHAT I BELIEVE

What do you believe? I'm a Christian and I believe that if you are someone who takes the Bible seriously, then you already care about climate change.

I know this might sound surprising. In the U.S., white evangelicals are less worried about climate change than any other group. But even when their objections are cloaked in religious-sounding language, it's not their theology that drives them. No, it's the political polarization and tribalism I talked about before. That's what's responsible for the partisan frames of many U.S. Christians, and it's those frames, not the Bible, that cause them to reject what science says about a changing climate.

Climate change disproportionately affects the poor, the hungry, and the sick, the very ones the Bible instructs us to care for and love. If you belong to any major world religion—or even if you don't—this probably speaks to you, too. Climate change amplifies hunger and poverty, increasing risks of resource scarcity that can in turn exacerbate political instability and even create or worsen refugee crises. Those most vulnerable to climate change are the same people who already suffer from malnutrition, food shortages, water scarcity, and disease. That's true here in North America as well as on the other side of the world.

Then there's pollution, biodiversity loss, habitat fragmentation, species extinction: climate change makes all those worse, too. Dominion is not the same thing as domination; the very word implies stewardship and sustainability, with more animals on the ark than people, so to speak. In fact, if Christians truly believe we've been given responsibility— "dominion"—over every living thing on this planet, as it says at the very beginning of Genesis, then we won't only objectively care about climate change. We will be at the front of the line demanding action because it's our God-given responsibility to do so. Failing to care about climate change is a failure to love. What is more Christian than to be good stewards of the planet and love our global neighbor as ourselves?

BE WHO YOU ARE

Bombarding people with more data, facts, and science isn't the key to convincing others of why climate change matters and how important and urgent it is that we fix it. Instead, when we're talking about contentious, politicized issues, study after study has shown that sharing our personal and lived experiences is far more compelling than reeling off distant facts. Connect *who we are* to *why we care*. Bond with someone over a value we already possess and share, one that is already near and dear to our hearts. Tell them why you care about climate change and why others might, too.

Contemporary theorists like George Marshall explain the power of this approach. Caring about and acting on climate change enables us to be an even more authentic version of who we already are, he says. Being a loving parent, or an avid conservationist, or a savvy businessperson, or a defender of your country, or a devout believer isn't only consistent with climate action. Such action gives you a new opportunity to better express, through your words and your actions, who and what you care about.

So now, when I'm not sure how someone will react to a conversation about climate change, I don't begin with that. Instead, I start by asking them about themselves. I share who I am, placing particular emphasis on those aspects that we have in common. I'm a Canadian; someone who lives in Texas and wants the best for it; someone who loves winter and snow and the outdoors and her child; and someone who's a Christian who believes that we humans have a God-given responsibility to care for our home, and for our sisters and our brothers and every living thing that shares it with us.

The World Evangelical Alliance represents 600 million people around the world. Since 2019, I've been honored to serve as their climate ambassador. They take climate change so seriously that their former secretary general Bishop Efraim Tendero was an official member of the Philippine delegation to the Paris climate conference in 2015.

A North American evangelical once asked Bishop Efraim why he cares about climate change, adding accusingly, "Don't you read the Scriptures?"

"Yes," he replied, "I do read the Scriptures. *That's exactly why I care.*"

3

WHO YOU ARE

"The environment is where we all meet, where we all have a mutual interest; it is the one thing all of us share. It is not only a mirror of ourselves, but a focusing lens on what we can become."

LADY BIRD JOHNSON

"How do I talk to my friends about this? They don't see it like I do."

NASA RESEARCHER AT KATHARINE'S TALK

I never used to talk about why I became a climate scientist. Getting personal about our work is something we scientists are trained not to do. Leaving our deeper motivations at the door of the lab serves us well as we analyze data and draw conclusions. But when it comes to explaining to others—and to ourselves—why it matters, we need more.

When I was nine years old, my family moved to Cali, Colombia, where my parents spent several years working at a bilingual school and helping out with a local church. We lived in a whitewashed, red-tiled row house in a lower- to middle-class area. We had electricity and running water most days, and we'd fill up the kerosene lamp and the bathtub for when we didn't. But we often spent the weekends visiting far-flung barrios where houses of mud bricks and tin clung to the side of a mountain, or remote villages that lay several hours up a rutted, puddle-filled dirt road. There, the toilet was simply a wall to stand behind—wearing a skirt made it a lot easier—and lunch consisted of rice, plantains, and the guinea pigs that were slowest to run.

Our vintage 1969 Ford Bronco would be packed with everyone who needed a ride there or back, plus my dad's telescope—an avid amateur astronomer, he'd often host impromptu viewings of the moon and Saturn's rings. If it had been raining, I'd be propped up in the driver's seat, one toe barely reaching the clutch, to steer while everyone else got out to push the truck through the mud.

Disasters hit us hard here in North America, but our recovery is cushioned by private and public services, from home insurance to disaster relief. In Colombia in the 1980s, life was challenging at the best of times: poverty, inequality, lack of clean water and health care; corruption and danger from the mafia, the guerrillas, the paramilitaries; reverberations from the atrocities of "La Violencia" in the 1950s still echoing through many rural areas. When disaster struck, it could be devastating. When rains came, entire neighborhoods were swept away. When drought hit, people starved.

Inspired by my dad's love of the night sky, I decided to study astrophysics when I returned home to attend university in Canada. But a course I serendipitously took on climate science yanked my focus from the stars to the Earth. I learned how climate change is making many of the risks faced by people in low income countries worse; that is why the U.S. military calls it a *threat multiplier*. Climate change affects our food, our water, even the air we breathe. It accelerates the destructive impacts of human expansion on natural ecosystems and it impacts our own health, our welfare, even our pocketbooks. And it exacerbates humanitarian crises: poverty, hunger, diseases, even political instability and the plight of refugees.

I learned how the vulnerabilities I saw in Colombia are reflected around the world and are being amplified by climate change. Just like the coronavirus pandemic, it's deepening the chasm between the haves and the have-nots, pushing many more into poverty. Whoever we are, wherever we live, disasters take whatever challenges we are already facing and make them worse.

As a Christian, I believe we're called to love others as we've been loved by God, and that means caring for those who are suffering—their physical needs and their well-being—which today are being exacerbated by climate impacts. How could I not want to do something about that?

That's why I became a climate scientist.

BEGIN WITH WHO YOU ARE

The first time I was invited to speak at a local church here in Texas, I decided the time had come to share more of my personal motivation, as uncomfortable as it might feel. After all, the reason I'm a climate scientist is *because* I'm a Christian. Maybe, I thought, just maybe a few of the people there might realize they cared about climate change for the same reasons I did.

It was a Wednesday night. The meeting was in one of the adult Bible study rooms, down a long hall with tan carpet. A group of about fifty interested people had gathered. I showed them the data revealing that yes, the planet is warming, and yes, humans are responsible. As I expanded on the impacts we were already experiencing in Texas, people listened and nodded along; they felt validated by what I had to say and it matched what they'd witnessed themselves. But then I took a deep breath, gathered up my courage, and for the first time ever, nervously launched into why I cared: the biblical mandate for stewardship and care for creation, the connection between climate change and poverty, and the Bible verses that directed my concern.

I was half expecting people to laugh; but instead, they seemed surprised. They recognized those Bible verses I was quoting and they lived by the same principles. And the questions I got afterward shifted: they were deeper, far more personal than I'd heard before. This audience *cared*. Why? Because we had connected over something fundamental and undeniable that we shared.

BRING FAITH INTO THE CONVERSATION

Tim Fullman is an ecologist who studies caribou in Alaska. He grew up in a conservative evangelical church in Southern California. Although he was encouraged in his love of science, certain topics—creationism and evolution—were only considered from one viewpoint. For a long time, his perspective on climate change could best be summarized as "I don't really know either way, but I've at least got some skeptical questions," but he hadn't really taken the time to look into the issue.

It was only later, when he was in grad school at the University of Florida studying geography, that he started to question why climate change was a "Christian issue." "This led me to ask a Christian mentor the same question," he says, "and his response surprised me. 'It's because this is the kind of thing that will lead to big sweeping regulations and a one-world government, as you read about in the Book of Revelation. It could be a sign of the end times.'"

As a Christian, Tim could understand why there was debate over evolution, but his mentor's reply left him with more questions than answers. This prompted him to look deeper. "I couldn't find anything that clearly contradicted climate change from reading the Bible," he says, "so I felt like I should take a more open look at the scientific evidence." When he did, it was enough to convince him. "I saw the potential alternative drivers of climate change—volcanoes, the Sun, the Earth's orbit—failing to fit the historic data," he says, "and human-produced greenhouse gases fitting it nicely."

One of the most frequent "Christian" arguments I hear is that God is in control, so humans can't affect something as big as the Earth. But this completely overlooks the role of human agency, the fact that the Bible explicitly states that God gave humans responsibility over the Earth. And Tim agrees. "Even though my ultimate hope and security is in heaven," he says, "I think we are called to stewardship and what we do here on the Earth really matters."

Tim now makes it a priority to communicate about climate change with others, especially with friends and family in the conservative sphere. Five years ago he would not have brought this topic up, but now he values helping other Christians see that this doesn't conflict with their faith. "Actually, caring about the climate and the environment is how we 'love the least of these,' and help other people," he reflects. His values didn't change; if anything, he developed a deeper understanding of them, one he can now communicate to those who share them.

If you're not a Rotarian or a person of faith, don't be discouraged. These are only two of many ways we can connect with others. Another effective point of connection is through a shared passion or interest.

START WITH WHAT YOU LIKE TO DO

Renée is a ski racer from Quebec. Like many of her high school class-mates, she cared about climate change but was starting to despair of being able to alter anything. "I felt overwhelmed," she says. "The prob-lem was so big, and a lot of people knew that climate change was an issue, but nobody was doing anything about it."

That all changed when she heard the words of young Swedish activ-ist Greta Thunberg, who inspired her to go to the big climate strike in downtown Ottawa in September 2019. "That sense of community made me think, 'I'm really not in this alone. A bunch of other people are fight-ing for the same goals,'" she said.

With her friend and fellow skier Julia, she joined Protect Our Win-ters (POW), an organization started by Jeremy Jones. He's an American snowboarder worried about the effects that climate change is having on the sport. Now she recruits other skiers to spread the message. To help, she's going to make helmet stickers, with the Protect Our Winters logo, out of recycled materials. "There's no waste coming from the stickers," she says, "and they can start a good conversation and get more people involved."

Renée has given presentations in class about Protect Our Winters and described ways her classmates can all minimize their impact on the environment. For Earth Day, she helped organize an assembly at her high school with a speaker from POW's Hot Planet/Cool Athletes out-reach program. She's realized that although hardly anyone ever changes their mind in the span of a conversation, helping people connect who they are to why they care makes a big difference, long-term. Now in university, she's decided to major in environmental studies.

TALK ABOUT WHAT YOU LOVE

Who else's interests are directly impacted by climate change? Well, birders, for one. The Audubon Society has put together over 140 million
observations by birders and scientists to create a compelling series of
maps showing where hundreds of bird species live today and how their
ranges will shift due to climate change. They estimate that two-thirds
of the birds in North America are at risk of climate-related extinction,
and iconic species may lose the meaning of their names—the Baltimore
oriole, for example, may no longer be native to Baltimore. The Audubon's conclusion is crystal-clear: "Birds will be forced to relocate to find
favorable homes. And they may not survive." So if you're a birder, you
have every reason to care about climate change.

If you enjoy fly fishing, you need to know that climate change is increasing the water temperature in streams, making fish such as salmon
and trout more susceptible to disease and parasites. Many streams are
snow-fed: warmer winters mean less snow, more rain, and earlier snow-
melt. This affects the timing of fish migration and lowers summer flow.
In Oregon and Idaho, it's estimated that up to 40 percent of salmon habitat could be lost before the end of the century due to warmer temperatures alone.

Then there are hunters. Ducks Unlimited says that conservation has
made great strides in restoring bird populations—but climate change
could undo all its gains. "Climate change poses a significant threat to
North America's waterfowl," they warn, "that could undermine achieve-
ments gained through more than 70 years of conservation work."

Snowmobilers might not seem like the most obvious group to be
concerned about climate change, yet a study I led for the U.S. Northeast
found that the snowmobile season in many states is already shrinking.
Across much of eastern North America, recreational snowmobilers are
essential to the wintertime economy of many small towns. Motels, restaurants, shops, and gas stations depend on their traffic. As climate continues to change, only the most northern locations will still have enough
snow on the ground on a regular basis to sustain the industry.

Warmer temperatures and heavier rain events are affecting the viability of many other outdoor sports, from golf to soccer. Record summer heat waves since their last stadium opened in 1994 have already prompted the Texas Rangers baseball team to build a whole *new* stadium. They put a roof on it and air-condition it to keep the fans and players cool. Pond hockey and outdoor ice rinks, a staple in many northern backyards and neighborhood parks, are getting less and less viable as winters warm.

Cities hosting the Winter Olympics have to worry about whether there will be snow on their mountains; Summer Olympic hosts like Tokyo are concerned about extreme heat. Biometeorologists like my friend Jenni Vanos from Arizona State University are being asked to map out marathon routes that keep athletes cool. Outdoor tennis competitions are taking longer breaks to minimize athletes' risk of heat exhaustion.

Even those of us who aren't into sports often look forward to a beach vacation. However, many beaches are already being eroded and submerged by rising seas. Half the world's sandy beaches could be gone by end-of-century, with many beachside economies—in Australia, Brazil, and the U.S. Gulf Coast region—along with them.

It's hard to think of *any* outdoor activity that isn't being affected by climate change.

BRING UP WHAT YOU GROW AND EAT

Gardeners are seeing plant hardiness zones shifting. Much of the U.S. has moved into a whole new zone in just twenty-five years. Plants are flowering and blooming earlier in the year, and invasive species are migrating into many regions. It's estimated invasive species have cost the global economy over a trillion dollars since 1970.

I'm part of the U.S. Department of the Interior's South Central Climate Adaptation Science Center (CASC). We work with landowners, farmers, and ecologists worried about species like fire ants and bindweed spreading across our region as it warms. In the Northeast CASC, scientist Bethany Bradley decided to be more proactive about it. She and her colleagues created brochures and a website to teach Northeastern gardeners which non-native plants to avoid: burning bush, Japanese honeysuckle, and, most of all, kudzu. During the 1930s, farmers in the Southeast were encouraged to plant kudzu to feed to their livestock. Lacking any natural predator, this woody plant soon became known as the "vine that ate the south." Thanks to warmer winters, over the last few decades kudzu has been spreading northward. It's even reached southern Ontario.

In 2016, the Garden Club of America released a statement expressing their concern over climate change, and their commitment to educating people about "causes and constructive responses." Each year now, they reach out to climate scientists like me to give them the latest update on what we've learned and how they can help.

Climate change also affects the food we grow. Greens like lettuce could face unfavorable growing conditions and more extreme weather events that make contaminants like E. coli easier to spread. And warmer temperatures could imperil the water supplies needed to grow bananas, the world's most popular fruit, as well as citrus fruits and olives. As temperatures warm, conditions become more favorable for many of the pests and diseases that already destroy between 20 to 40 percent of the world's crops every year.

It's not only food, either; your favorite beverage might be impacted, too. Warmer temperatures and higher levels of carbon dioxide in the atmosphere affect the composition and fermentation of grapes. This is already happening in iconic wine-growing areas of France, from Languedoc to Bordeaux; and it's affecting beer, too. Warmer temperatures have decreased the yield of hops and altered beer quality in the Czech Republic, which is famous for its lager. In 2015, twenty-four breweries, including Guinness and New Belgium, signed a Brewery Climate Declaration, calling attention to the risks climate change poses to their industry. And some organizations are going further: SAB Miller, the parent company of familiar names like Grolsch and Miller Genuine Draft, is experimenting with cassava root as a replacement for barley malt in warmer growing areas.

Prefer hot chocolate or coffee? Shifts in rainfall patterns are already affecting cacao harvests, and warmer temperatures increase evapotranspiration, essentially squeezing the water out of the soil and plants that produce chocolate. Coffee giants from Nespresso to Lavazza are concerned about climate change impacts on the millions of small shareholder farmers who grow beans around the world. They are launching programs to build resilience and even provide customized insurance to cover farmers' climate-related losses due to disease, mold, drought, and more.

By now, you might see a pattern, and some ways into conversation through things that people love and you do, too. Do you like gardening, beaches, or birds? Beer, coffee, or wine? Outdoor experiences, sports, or activities with your family? If so, then you have something to talk about.

GO BEYOND THE SCIENCE

Many scientists have particular difficulty figuring out how to connect with nonscientists over climate change. I think that's at least in part because being a scientist isn't just a job or a career for most of us. It's more like a vocation or a calling, a fascination and a lifelong quest for knowledge.

We scientists connect with one another over that shared love of science without even consciously recognizing we're doing so. But when you think, breathe, and live science, it can be hard to decipher what *else* you care about. As a result, when giving a talk at a university or to an academic audience, I typically field a dozen or more questions that are all variants of "How can I connect with anyone over anything other than science?"

After a Christmas event in Washington, D.C., a NASA researcher approached me, overwhelmed with concern about how climate change is affecting our world and how he sees this in the data, firsthand. "But how do I talk to my friends about this?" he asked. "They don't see it like I do." I asked him what he enjoys doing with his friends, and he said they love getting together to cook. He comes from South America, one of many regions where climate change is increasing the risk of drought as well as affecting specialty crops and people's livelihoods. All of these would be relevant things to bring up at his next dinner party, I said.

A young woman who worked for a science organization sought me out, wondering how to broach the topic of climate change with her grandmother. "What do you like to do together?" I asked. "Knit," she replied. I love knitting, so the answer came easily. I suggested she look up the "warming stripes" for the place where her grandmother grew up. Warming stripes are a visual representation of how temperature has changed in any given location over time. The creation of British climate scientist Ed Hawkins, warming stripes consist of one stripe for each year. If it was a colder year, it's blue. A normal year is white. A warmer year is pink, and a hotter year is red. People have turned the warming stripes into knitting patterns that start off mostly blue at one end—usually around the late 1800s or early 1900s—and then shift to bright red as they head toward the 2020s. "What if you knit a scarf together," I said, "and ask your grandma to tell you about memorable years, and how she's seen conditions change over her lifetime?"

After speaking at a university in California, a scientist told me he'd been reaching out to local churches. He was convinced, as I am, that getting churches on board is part of the solution—but he hadn't been getting any traction. What should he do?

"Start with the denomination or the type of congregation you have most in common with," I suggested. "Where do you attend?"

"Oh, I'm an atheist," he said blithely.

"In that case, stop!" I replied. "You are not the right person to have that conversation; that's why it's been falling flat.

"Instead, connect with people whose values you genuinely share. So, what do you enjoy doing?"

"Well, science," he said, as a scientist always will.

"Yes," I said, "of course. But what else do you do? Do you hike or run, sail or surf; are you a member of the Rotary Club or any community organizations?"

"No," he replied, "and no, and no."

After going through about ten different options, something finally occurred to him.

"Well, I am a diver," he said uncertainly, not convinced it mattered. But when I smiled encouragingly and nodded, he warmed to his theme. "I dive a lot, and I've been doing it for a long time. In fact, I hold a few records for very deep dives."

"Perfect!" I said. "That's exactly the community you can reach out to. Oceans are being affected even more than land areas due to warming and acidification, but we humans often don't realize it because we don't live in the oceans. You could approach diving instructors and schools and clubs in your area and offer to educate divers on how climate change is affecting the oceans and marine life and what they can do to help. They are much more likely to listen to you than to me. You are a diver, so you understand."

WHAT ABOUT YOU?

At this point, if you're a food or drink aficionado, a crafter, an outdoor enthusiast, or an ocean-lover, you may be starting to see a few ways you hadn't explored before that you can build on to connect with others. But if you're not one of those, and not a particularly religious person either— or even if you are but that's not a part of your world that you're wanting to have conversations in—then you might be thinking, "That's all well and fine for *you*, Katharine, but this still doesn't help *me*."

It will, though: because being a person of faith, or an athlete, a gardener, or a wine connoisseur, someone who knits, or someone who loves snow, aren't the only frames that work. Depending on who you are, whom you're talking to, and what you both care about, there are all kinds of approaches that will work for you. The only condition is that you have to be genuine. Don't pretend to be something you're not.

If you're still struggling with this concept and how it applies to you, ask yourself a few questions. Where do you live? What or whom do you love? What activities do you enjoy doing? What do you do for work?

Do you come from any particular culture, place, or faith tradition? And perhaps most importantly, what are you passionate about?

It could be the simple fact that you and the other person are both parents, or you live in the same place. Or you could bond over the fact you work in the same industry or business, or you enjoy the same types of activities. You might be an avid online gamer, like oceanographer Henri Drake, who runs a Twitch channel to talk about climate change in his spare time. You might be a hockey player like climate modeller Gabe Vecchi, who tracks how outdoor skating days are declining around Princeton, where he lives, and uses that information to talk to his teammates about climate change. How you connect with others doesn't have to fit any mold, example, or pattern. Whoever you are, *you* are the perfect person to talk about climate change with others who share your interests and concerns.

WE ALREADY CARE

When you find out what people care about, and connect climate impacts directly to the values people have, they can see that caring about climate change is already integral to who they are. Parents care about their kids' health, and future. Residents of a city or region care about their water supply and their local economy. Outdoor enthusiasts care about the abundance of fish, birds, wildlife, or snow, each of which requires a healthy environment and a thriving ecosystem. Military and defense experts, from Pentagon employees to four-star generals, are very worried about climate change and its potential to multiply resource scarcity and security threats around the world.

To put it another way, none of us cares about climate change because a two-degree or three-degree or even a four-degree increase in the average temperature of the planet matters to us personally. I don't even care for that reason, and I'm a climate scientist. We care because the cascade of events triggered by that warming affects everything we *already* care about: where we get the food that we eat and how much it costs; how clean or dirty the air that we breathe is; the economy and national security; hunger, disease, and poverty across the planet; the future of civilization as we know it. We've woven a million reasons why we already care about climate change into the very fabric of our society. We just haven't fully realized it yet.

Stop for a second and take a breath. That came from our planet. Then think about how every single resource we use in our lives is provided by the Earth, from the water we drink and the food we consume to the materials we use to build our houses and our clothes and our phones. All of those resources are gifts from our home, planet Earth.

This is why, to care about a changing climate, we don't have to change anyone's values or try to transform them into anything other than who they already are. We just need to be people who want this planet to continue to be a safe, hospitable home for us all.

And to share this message effectively, we need to bring our hearts to the table, not just our heads.

WHY FACTS MATTER–AND WHY THEY ARE NOT ENOUGH

4

THE FACTS ARE THE FACTS

"The truth does not change according to our ability to stomach it."

FLANNERY O'CONNOR

"What science do you and Biden subscribe to that makes this change human-caused but all the others totally natural?"

AN ENGINEER ON LINKEDIN, COMMENTING ON KATHARINE'S POST

"Last month we visited the Creation Museum," our friend Mark said brightly. We were catching up over coffee after we hadn't seen each other in a while. My husband nudged me, a silent warning not to say anything impolite: he'd already heard my opinion of the museum's statues of humans alongside animatronic dinosaurs a few too many times.

The special exhibit on climate change was particularly interesting, Mark said. "A World War Two airplane that crashed over Greenland was found buried under more than two hundred feet of snow," he enthused, holding up a photo on his phone of the museum's diorama. "So those ice cores from Greenland that scientists claim show what climate has been like for hundreds of thousands of years can't be real. They only show the last few decades!"

As with nearly all scientific-sounding objections to climate change, this argument seemed legitimate on the surface. Propaganda works best by wrapping falsehoods around something true (in this case, a real plane crash in Greenland). But this one was easy to debunk. Snowfall rates on Greenland's coast are high; that's why the plane got buried so quickly. Inland, though, where scientists drill the ice cores, there's a lot less snow. As more snow falls on top of it, older snow gets compressed and compacted into ice, eventually sealing off the bubbles of air in it. That's how a hundred-thousand-year-old ice core ends up being about three kilometers deep. The bubbles in it serve as records of past changes in heat-trapping gases, dust, air temperature proxies, and more. They're carefully checked against other records from around the world to make sure they are properly calibrated by year. There's no mistake: ice cores do tell us about ancient climate, and they also show us how unusual today's climate is.

––––––

Every day I'm bombarded with objections to the evidence for human-caused climate change. Most of them sound respectably scientific, like "climate changes all the time; humans have nothing to do with it." It's "the Sun," or "volcanoes," or "cosmic rays" that are making it happen, or "it's not even warming," some argue.

You've probably had to face these objections, too. Maybe they come from a family member, quoting their favorite politician; a colleague, citing a questionable blog; or something you run across yourself on social media that you know can't be true, but you can't quite put your finger on why it isn't. Scientific-sounding objections are the number one type of objection we hear when people want to argue about climate change. Fully half the questions I receive on Facebook are some version of, "Someone I know posted this article. I know it can't possibly be true. But can you explain why?"

Scientists call these "zombie arguments." They just won't die, no matter how often or how thoroughly they're debunked. And because they won't die, it's clear that, when it comes to climate change, you *do* have to be able to talk about some of the science. But what you *don't* have to do is follow it down the rabbit hole. And you don't have to be a climate scientist, either. The basics are extremely simple, the objections are very common and easily answered, and it makes sense to have a short response at hand.

Not only that, but as I mentioned before, most scientific-sounding objections are really just a thin smoke screen for the real problems. Climate denial originates in political polarization and identity, fueled by the mistaken belief that its impacts don't matter to us and there's nothing constructive or even tolerable we can do to fix it. Again, this isn't only a U.S. problem: an analysis of people across fifty-six countries found that political affiliation and ideology was a much stronger indicator of their opinions on climate change than their education, their life experiences, or even their values.

But as Ronald Reagan said, "If you're explaining, you're losing." So the key when these zombie arguments surface is to have an answer, but to keep it short. Acknowledge the objection, and provide a brief response. Then pivot promptly to connecting over shared values rather than divisive arguments, from the heart rather than the head. Here's what you need to know about the science so you can do that.

THE EXPLANATION IS SIMPLE

The Earth's climate is complex. Understanding what we humans are doing to it isn't. Think of it this way: The Earth is wrapped in a natural blanket of heat-trapping gases. Most of the Sun's energy goes right through this blanket, as it does through a window, heating the Earth.* The Earth absorbs the Sun's energy. It warms up, and gives off heat energy. The blanket traps the heat energy, keeping the Earth around 33°C or 60°F warmer than it would be otherwise. In fact, if we didn't have this blanket, the planet would be a frozen ball of ice.

So if this blanket is natural, and it's responsible for the fact that there's life on Earth, what's the problem? The problem is that whenever we dig coal, oil, or natural gas out of the ground and burn it, we release carbon

---

*The Sun is very hot, so most of the energy it gives off is in the shorter visible and infrared wavelengths. Heat-trapping gases don't absorb much energy at these wavelengths. In contrast, the Earth is a lot cooler so most of the energy it gives off is in the longer infrared wavelengths. That's exactly where heat-trapping or "greenhouse" gases absorb the most heat.

dioxide or CO_2 into the atmosphere—carbon that would not naturally reach the atmosphere for millions of years. And carbon dioxide is one of the main gases that make up our heat-trapping blanket. Heat-trapping gases also come from deforestation, agriculture, and waste. Hundreds of years' worth of carbon dioxide, methane, and nitrous oxide emissions have artificially increased the thickness of that natural blanket. You'd overheat if someone replaced your perfect blanket with a thicker one you didn't need. In the same way, the Earth is also heating up.

I know that each time a bad blizzard or a cold spell engulfs us, you might think, "I sure could use some global warming right now." But as comedian Stephen Colbert tweeted sarcastically in 2014, "Global warming isn't real because I was cold today. Also great news: World hunger is over because I just ate."

What's happening in one place on one day, or even one year, doesn't invalidate the long-term warming of the entire planet. The truth is that no matter how cold or hot it is today, no matter what season it is, each successive decade is breaking new ground as the warmest on record at the global scale.

THE SCIENCE IS VERY OLD

French mathematician and scientist Joseph Fourier was the first to identify our planet's natural blanket, in the 1820s. In 1856, Eunice Foote, an amateur scientist from New York—in a paper presented on her behalf at the annual meeting of the American Association for the Advancement of Science (AAAS)—proposed that if carbon dioxide levels in the atmosphere were higher, the planet would be much warmer. The next year she became one of the first women to read her own work at the AAAS annual meeting, and it's a conference I still attend today.

At the same time in the U.K., an Irish scientist named John Tyndall, with the greater benefits and resources the educational system provided to the men of that era, was inventing the delicate scientific instruments needed to measure precisely how much heat was absorbed by carbon dioxide and "coal gas" (primarily methane). By the late 1800s, scientists could calculate exactly how much the planet would warm as carbon dioxide in the atmosphere increased. By the 1930s, British engineer Guy Callendar could actually measure how temperature had changed since the 1880s due to burning fossil fuels.

By 1965, White House science advisors were confident enough that climate change was real, humans were causing it, and the impacts were serious to formally warn a U.S. president, Lyndon Johnson, of the dangers increasing atmospheric carbon dioxide posed to Earth's climate. "Within a few short centuries, we are returning to the air a significant part of the carbon that was extracted by plants and buried in the sediments during half a billion years," scientists wrote in their report. "By the year 2000 the increase in CO_2 will be close to 25 percent [relative to pre-industrial times]. This may be sufficient to produce measurable and perhaps marked changes in climate." As with many early projections, this proved remarkably accurate.

In 1987, *TIME* magazine put a burning planet in a greenhouse on its cover. The next year, NASA scientist Jim Hansen testified to Congress that global warming was real. That same year, the United Nations Intergovernmental Panel on Climate Change (IPCC) was formed, and in 1990 they released the first of their now six exhaustive and ever-expanding assessment reports. These reports document everything scientists know about how climate is changing and how it will impact our world.

The Earth Summit was held in Rio de Janeiro in 1992. Nearly every country in the world, including the U.S., signed the resulting U.N. Framework Convention on Climate Change. In it, they agreed to "prevent dangerous anthropogenic [i.e., human] interference with the climate system"—but they couldn't agree on what's considered dangerous until they'd met twenty-one more times. Not until the Paris climate conference in 2015 was the world able to agree to keep "global temperature rise this century well below 2 degrees Celsius above pre-industrial levels and to pursue efforts to limit the temperature increase even further to 1.5 degrees Celsius." This is what's known as the Paris Agreement.

In 2016, thirty-one scientific organizations sent a letter to Congress: "We, as leaders of major scientific organizations, write to remind you of the consensus scientific view of climate change," they wrote. "Observations throughout the world make it clear that climate change is occurring, and rigorous scientific research concludes that the greenhouse gases emitted by human activities are the primary driver. This conclusion is based on multiple independent lines of evidence and the vast body of peer-reviewed science." By 2020, eighteen scientific societies in the United States, from the American Geophysical Union to the American Medical Association, had issued official statements on climate change and one hundred and ninety-eight scientific organizations worldwide had formally stated that climate change has been caused by humans. That's how sure we scientists are.

RULING OUT THE OTHER SUSPECTS

Despite this overwhelming history of scientific understanding and global consensus, I still hear objections to the science every day. They don't come from other scientists, but from people asking, "Don't you know it's been warmer before? Humans weren't causing climate change millions of years ago. So why would you think we're responsible now?"

What many don't realize is that scientists don't automatically assume it's human-caused without checking any other options first. Just as a responsible and knowledgeable physician would first rule out all common causes of a persistent low-grade fever—infection? autoimmune disease? cancer?—so, too, have scientists rigorously examined and tested all other reasons why climate could be changing naturally. That's why we're so sure it's humans this time: because natural factors all have an alibi. Here's how we know.

IS IT THE SUN?

The Sun is the first and biggest "natural suspect." That's because the Earth gets nearly all of its energy from our nearest star. The Sun's brightness fluctuates over time—a lot, over astronomical timescales, and a little, over human timescales. When energy from the Sun increases, the planet warms up slightly. This is similar to how a room brightens when you turn up the dimmer on a lamp. When the Sun's energy decreases over decades to centuries, the Earth gets slightly cooler.

During some past cooler periods, such as the little Ice Age from the 1400s to the 1800s, the Sun's energy was slightly below average. In the northern hemisphere, winter temperatures dropped by 1 to 2°C (1.8 to 3.6°F), for several centuries. In London, it became common for the Thames to freeze solid, and Londoners held "frost fairs" out on the ice. Some winters, so much sea ice encircled Iceland that the island was unreachable by shipping.

For today's warming to be due to the Sun, though, its energy would have to be increasing—and it's not. Since the 1970s, satellite radiometer data show that the Sun's energy has been decreasing. So if the Sun were controlling our climate right now, we'd be getting cooler. Instead, Earth's temperature continues to increase. The Sun has an alibi.

WHAT ABOUT VOLCANOES?

There's a popular myth that one volcanic eruption produces ten times more carbon pollution then all 8 billion of us humans put together. In reality, though, volcanic eruptions don't warm the Earth: they mainly cool it. When volcanoes erupt, they expel vast clouds of sulfur dioxide into the atmosphere. These molecules combine with water vapor to create sulfuric acid "aerosols." These aerosol particles absorb some of the Sun's rays and reflect some back to space, acting like an umbrella to cool the Earth.

One of the largest eruptions in human history occurred in 1815 when Mount Tambora, on the Indonesian island of Sumbawa, spewed out over 60 megatons of sulfur dioxide. For three years afterward, global temperatures dropped noticeably. This eruption has been blamed for everything from massive crop failures across the northeastern United States to famine in Europe to the disruption of the Southeast Asian monsoon season. The year 1816 became known as "the year without a summer." Mary Shelley spent much of that dreary summer indoors in Switzerland, and the gloom inspired her to write *Frankenstein* and likely *The Last Man*, an apocalyptic and eerily prescient novel that covers plague, climate refugees, and yes, reports of a "black sun" that leads to mass panic. The Mount Tambora eruption spawned outbreaks of typhus across Europe and cholera in India, and even the heavy rainfall and flooding that was ultimately responsible for Napoleon's defeat at Waterloo. But it didn't heat the Earth.

It's true that in geologically active regions such as Iceland, Sicily, and Yellowstone National Park, heat-trapping gases seep from the Earth's crust into the atmosphere. These do have a warming effect. A small amount of heat-trapping gases are released by eruptions as well. But natural geologic emissions amount to around 1 percent of the carbon dioxide and less than 15 percent of the methane that human activities contribute to the atmosphere every year. All geologic emissions put together are equivalent to the human emissions of about three midsized U.S. states, such as Virginia, Tennessee, and Oklahoma. That's minimal compared to how much carbon humans have been pumping into the atmosphere every year. So no, volcanoes aren't causing the planet to warm, either.

COULD IT BE ORBITAL CYCLES?

A third legitimate climate change suspect is orbital cycles. These are caused by periodic variations in the Earth's orbit around the Sun. Orbital cycles are responsible for the ice ages, or glacial maxima, that our planet has experienced in the distant past. They're also responsible for the warm interglacial periods, such as the one the Earth has been experiencing for the last twelve thousand years or so. But are they responsible for the warming over the past one and a half centuries? No, and here's why.

In the early 1800s, scientists realized that large sheets of ice once covered Europe and North America. For a long time, though, they didn't know what had caused these ice ages. It wasn't until the 1920s that a brilliant young Serbian civil engineer and mathematician named Milutin Milanković figured it out. Over time, the varying gravitational pull of the larger planets stretches the Earth's orbit around the Sun from a circle to an ellipse and back again. The axis of Earth's own rotation also wobbles like a top. Charting six hundred thousand years of these variations by hand, he discovered that their cumulative effect creates cycles of about one hundred thousand years—the same length as the longest ice age cycles. These cycles alter how sunlight falls on the Earth, which in turn triggers the growth and retreat of the ice sheets.

The last major glacial maximum was twenty thousand years ago, so people often wonder if the planet is warming today because it's still recovering from the most recent ice age. Sadly, it isn't. The warming due to orbital cycles peaked about six to eight thousand years ago, in the early days of human civilization. At that point, the Earth's temperature started very gradually decreasing. According to orbital cycles, the next major glacial maximum was due to begin in about fifteen hundred years from now. *Was* due, that is; because about a hundred and fifty years ago, the planet started getting warmer instead.

COULD IT BE A NATURAL CYCLE?

If the Sun, volcanoes, and orbital cycles cannot be the cause of the current warming, that leaves just one more main natural suspect, and it's the one that's most commonly invoked: natural cycles. But what exactly *is* a natural cycle, and how does it warm or cool the planet?

Natural cycles can't create heat out of nothing. Rather, they help distribute energy around the planet by moving heat between the ocean and the atmosphere, or from east to west, and back again. They warm one part of the planet while simultaneously cooling another.

Some of the most well-known cycles even have names you might have heard of, like El Niño. During an El Niño episode, such as occurred in 2015, ocean temperatures off the coast of Peru and westward across the Pacific are warmer than average. As a result, the ocean releases heat into the atmosphere, which slightly raises the average global air temperature. It also typically brings drier-than-usual conditions to Australia and India, and wetter-than-usual conditions to the southern U.S.

In contrast, during a La Niña episode, such as occurred in 2020, cooler-than-average waters in the tropical Pacific absorb more heat from the atmosphere. Average global temperature drops slightly, flipping the pattern. It also brings wetter conditions to southeast Asia and much of Australia and drier conditions to the southern U.S.

During the so-called Medieval Warm Period, temperatures over the North Atlantic were about half a degree to one degree Celsius warmer than average for several centuries. This helped the Vikings to settle Greenland and reach northeastern Canada. Why "so-called"? Because it depends on your perspective: in Siberia during that time, it was actually the Medieval Cold Period. Temperatures over Siberia were colder than average, by about the same amount that temperatures over the North Atlantic were warmer. That's what a natural cycle looks like.

Today, however, the entire planet is warming, particularly the oceans; so it isn't just a natural cycle moving heat around. Over the last fifty years, the oceans have absorbed more than 90 percent of the heat being trapped inside the climate system. This means ocean heat content has increased about fifteen times more than that of the Earth's atmosphere, land surface, and cryosphere (the Earth's total ice and snow) combined.

Using the change in ocean heat content as a measure of climate change is far more accurate than tracking changes in air temperature. There's little year-to-year variability when you're looking at the steady increase in the heat content of the entire climate system rather than just one part of it, the atmosphere. So why do we hear so much about the increase in global air temperature, and not the ocean? Once again, it's because of our perspective. What's happening in the ocean—warming, acidification, and more—is even bigger and more alarming than what's happening on land. If we lived underwater, we'd realize that.

HUMANS ARE RESPONSIBLE

The bottom line is this: scientists have known since the 1850s that carbon dioxide traps heat. It's been building up in the atmosphere from all the coal, oil, and gas we've burned since the start of the Industrial Revolution to generate electricity, heat our homes, power our factories, and, eventually, run our cars, ships, and planes. Dozens of studies indicate that the most likely amount of warming humans are responsible for is more than 100 percent. How could it be more than 100 percent? Because according to natural factors, the planet should be cooling, not warming. We are the cause of *all* of the observed warming—and then some.

The planet has experienced warmer and colder times before. But as far back as we can look—and through paleoclimate records, we're able to look back millions of years—our present-day situation is unprecedented. The development of agriculture, with its large-scale deforestation and growing herds of methane-belching cows and other ruminants, was already likely enough to stave off the next ice age and stabilize climate. And that was a good thing: we don't want an ice-covered planet. But the Industrial Revolution kicked climate change into overdrive, and now the situation is dire.

Before the dawn of the Industrial Revolution, carbon dioxide levels in the atmosphere averaged around 280 parts per million (ppm), according to ice core records. Now, carbon dioxide levels are already more than 420 ppm, a 50 percent jump. The last time carbon dioxide levels in the atmosphere were this high was most likely over 15 million years ago. And the last time climate warmed at a similar pace to today was some 55 million years ago, during what scientists term the Paleocene-Eocene Thermal Maximum. That's when global temperatures rose by 5 to 8°C (9–14°F) over about one hundred thousand years, and sea level was over sixty meters (two hundred feet) higher than today. Scientists consider that period to be one of the most extreme examples of natural climate change on record. Today, it's estimated that we are currently emitting carbon into the atmosphere at *ten times* the pace of the natural emissions that drove this previous change.

What *is* the best temperature for humans? Neither hotter nor colder: it's the Goldilocks temperature we've had up until now. That's the temperature during which human civilization developed. It's the temperature during which we allocated our water resources, designed and built our infrastructure, and parceled out our agricultural land. These are the conditions under which we have developed our socioeconomic systems, outlined our political boundaries, and staked our ownership of natural resources.

We don't want another ice age, but today we've left it far behind. We are heading way too quickly in the opposite direction. Our climate isn't changing right now because of the Sun, or volcanoes, or natural cycles. Our change is human-caused. We humans are conducting a truly unprecedented experiment with the only home we have.

5

THE PROBLEM WITH FACTS

"We are so locked into our political identities that there is virtually no candidate, no information, no condition that can force us to change our minds."

EZRA KLEIN, *WHY WE'RE POLARIZED*

"I'd like to agree with you. But if I agree with you, I have to agree with Al Gore, and I could never do that."

FARMER SPEAKING TO KATHARINE

"Yes they are!"

"No they're *not*!"

At the podium was a meteorologist from the National Oceanic and Atmospheric Administration (NOAA) hurricane division. He was showing data indicating that hurricane frequency had not changed, long-term. Usually congenial, the meteorologist was clearly getting hot under the collar because shouting at him from the other side of the platform was a scientist from a NOAA research lab. This scientist's presentation, given immediately before, had showed exactly the opposite.

It was the annual meeting of the American Meteorological Society in January 2006. The record-breaking 2005 hurricane season had just ended. There'd been so many storms that, for the first time since the U.S. began to name Atlantic hurricanes in 1953, scientists had reached the end of the alphabet and had to wrap all the way around and begin with Greek letters instead.* The eleventh named storm of 2005, Hurricane Katrina, was the most expensive tropical storm on record. It caused over eighteen hundred deaths, $125 billion in damages, and countless personal losses for the inhabitants of New Orleans and the surrounding area. So it makes sense that scientists were wondering: where does climate change come in?

I was mesmerized. I'd never seen scientists spar like this before. Scientific disagreements are usually conducted via penned barbs and sarcastic one-way remarks—not face-to-face confrontations. But even though the two scientists were red-faced and loud-voiced, no fists flew, and no personal insults were hurled. Data was the only weapon they used; and eventually, data resolved the argument.

Today scientists know that the *overall* number of hurricanes isn't increasing, but the number of *strong* hurricanes is. We know why, too: warmer ocean waters fuel bigger, stronger storms and cause them to intensify faster. It turned out both scientists were correct; they were just looking at the data from different angles. Now we have a clearer picture of the whole. In the world of science, facts usually do win the day.

DEBUNKING FAKE NEWS

I spend a lot of time debunking science myths on social media. There, unfortunately, the same rules don't apply. In 2020, when wildfires destroyed record areas of the western U.S., people were claiming that climate change had nothing to do with it. "It's all arson," they argued, "or lack of forest management; or maybe even the fires aren't even real. After all, just look at Canada on this map—there's no fires there!"**

"I was on your Twitter," said my science teacher dad, who has the good sense not to have a Twitter account. "Why do you bother responding to those idiots?"

---

*In 2005 there were twenty-seven named storms, ending with Zeta. In 2020 there were thirty named storms, ending with Hurricane Iota, a category 5 storm in mid-November.

**Of course there *were* wildfires in western Canada, but maps based on U.S. federal data only show U.S. wildfires.

I told him it's important to counter disinformation. When these arguments occur publicly, there are others listening in who need to know that, as scientists, we have heard these zombie objections many times and we have good responses to them.

But in another sense, my dad was right. The positive feedback I've received for trying to respond with accurate data is sparse, though valued. "You dragged my sorry denier's ass to the truth," confessed one fellow Christian in a memorable tweet, and "You changed my dad's mind—thank you!" said another, privately. But such highlights are usually overwhelmed by the negative attacks from those who are unable to look at, let alone consider, something they see as a challenge to their identity.

"There's no convincing proof climate change is man-made," said one man. I replied with a thread I'd created earlier that systematically goes through each natural cause of climate change and shows why it can't be responsible for the current warming. "Insults are not proof," he responded, "keep trying, 'Professor'"—making it clear that he perceives facts to be attacks and expertise to be legitimate only if he agrees with what you say.

Of all confrontational people I've responded to with carefully marshaled and fully cited science, only a tiny handful have ever taken the time to engage in a thoughtful and honest way. So why aren't all these facts working to change people's minds? And if the facts are more accessible today than ever, why are so many people getting them so wrong?

PICKING YOUR OWN FACTS

Often, like my dad said, we think it's because people who disagree with us are idiots. There's even a book about it: *I'm Right and You're an Idiot*, by Canadian public relations expert Jim Hoggan. But his subtitle, *The Toxic State of Public Discourse and How to Clean it Up*, explains how the first assumption we jump to is a big part of the problem.

The vast majority of us understand that science and facts explain the way the world works, and that we ignore them at our peril. We all know that if someone says, "gravity isn't real" and steps off a cliff, they're going down whether they "believe" in it or not. So it's reasonable that individuals or institutions that want to change the way people think apply the *knowledge deficit model*: the idea that, if people disagree with some fact or scientific explanation, it's because they don't know enough. If that's true, then by implication, more information—better education, clearer explanations—will prevent people from making misleading claims about climate change.

This approach can work if we're talking about issues that don't have any moral or political baggage attached to them, like black holes or insect behavior. It can also work if we're talking about an issue that doesn't require immediate action, like astronomers' warning of a comet that might approach the Earth too close for comfort a century from now. But when politics, ideology, identity, and morality get tangled up in science—when our frames, as George Lakoff calls them, get in the way—then all bets are off. And what if that science implies that urgent and widespread action is needed? That's when the gloves come off, too.

Social scientist Dan Kahan has developed a measure he called "ordinary science intelligence." It measures how capable people are of understanding data, statistics, probabilities, and scientific results.* He then asked people whether they agreed that climate was changing due to human activities. He found only a weak correlation between science intelligence and the chance of a positive response. People with the lowest science intelligence had about a 35 percent chance of answering yes. People with the highest science intelligence had about a 60 percent chance of answering yes. But if he divided people by political affiliation, he found something quite different. Of these scientifically savvy people who identified as liberal Democrats, more than 90 percent were likely to answer the question positively. If they identified as conservative Republicans they had a greater than 90 percent chance of answering it *negatively*.

---

*This measure refers specifically to general scientific knowledge and analytical capabilities, not climate-specific knowledge. People who score high on climate-specific knowledge *do* tend to be more concerned about climate change.

Disturbingly, Kahan and colleagues found that "people with the highest degree of scientific literacy [that is, who were best able to understand science] were not most concerned about climate change. Rather, they were the ones among whom cultural polarization was greatest." In other words, the more broadly scientifically literate you are, the *more*, not less, likely it is that your political identity dictates your opinions on polarized issues like climate change.

Although Kahan's work was done in the U.S., a more recent study of people in sixty-four different countries found a similar trend. In other developed countries, education tends to make people more concerned about climate change, but this effect was noticeably diminished if they also identified as politically conservative. In developing countries and emerging economies, on the other hand, education made *everyone* more concerned about climate change regardless of their political affiliation.

It turns out that being better able to handle quantitative information and understand science in general doesn't make you more accepting of thorny, politically polarized scientific topics with moral implications that require a response; it just makes you better able to cherry-pick the information you need to validate what you already believe. For most of us, when it comes to climate change, we already have opinions. And the smarter we are, the harder we'll try to out-argue the devil if he disagrees with us. The term for this is *motivated reasoning*: an emotionally driven process of selecting and processing information with the goal of confirming what you already believe rather than informing your opinions or perspective. Which may beg the question—why did you pick up this book?

MOTIVATED REASONING

Basing our opinions and judgments on reason rather than emotion is the lofty goal laid out by Greek philosophers. It continues to be pursued by scientists today. But Plato might be disappointed to learn that modern psychology strongly suggests that when it comes to making up our minds about something, emotions usually come first and reason second. If we've already formed our opinions, more information will get filtered through those pre-existing frames. And the more closely that frame is tied to our sense of what makes us a good person, the more tightly we'll cling to it and let potentially opposing facts pass us by. As Jonathan Haidt explains in *The Righteous Mind*, "People make moral judgements quickly and emotionally. . . . We do moral reasoning not to reconstruct the actual reasons why we ourselves came to a judgement: we reason to find the best possible reasons why somebody else ought to join us in our judgement."

You might be thinking this primarily applies to people who don't care about climate change, but it actually cuts both ways. If we're worried about climate change, when we hear or see or read more bad news, it confirms what we already believe. The news essentially tells us, "See? You're right! This is real, and it's bad! It isn't just you saying this. It's these NASA scientists, and by extension, the Greenland ice sheet itself." This confirms what we already thought and reinforces our conviction that we are good people for thinking that. By extension, it also affirms that those who disagree with us are wrong.

When we want to believe something, psychologist Thomas Gilovich says, we ask ourselves "*Can* I believe it?" and we search for supporting evidence. When we don't want to believe something, we ask "*Must* I believe it?" and we search for contrary evidence. We *all* engage in this type of motivated reasoning when our identity is on the line, even when the stakes are relatively low. The smarter we are, the better we are—or so the social science warns us—at using motivated reasoning to defend our opinions and preserve our self-worth and identity.

For example, I grew up with the idea that you do not waste food. So if there isn't mold on it, I eat it (and if it's cheese, I just scrape the mold off and *still* eat it). My husband grew up with the belief that old food is bad food, and it will make you sick. And between two academics, it turns out you can find peer-reviewed articles supporting both eating food past its deadline and throwing it out. But there is a lot of motivated reasoning involved in our discussion, because "Waste not, want not" is part of my identity. On his part, whenever I feel sick, the first thing he asks suspiciously is "Did you eat those leftovers in the fridge from last week?"

Sometimes, though, all of us engage in motivated reasoning with higher stakes. In bigger decisions regarding parenting or religion, for example, our strong emotional attachment to a given position causes us to hold to our pre-existing opinion in the face of significant opposition and even solid fact. We will use all the intelligence we have to show why we're right, rather than admit we're wrong.

I fight this tendency in myself as a scientist, all the time. When you've invested years of work into something, it can be scary to imagine it might be wrong or off base. That's why I decided to join an international team a few years ago to re-analyze a handful of recent scientific studies that concluded that either the planet was not warming or that humans were not responsible. What if one of them had a point?

Our team leader, the Norwegian climate scientist Rasmus Benestad, collected the studies. There were thirty-eight in all, compared to the thousands that have been published over the same decade showing that the planet *is* warming and humans *are* responsible. He dismantled and re-calculated every single one of those studies from scratch. The rest of us followed behind, checking his work. And what we concluded was astounding. In every single analysis, we found evidence of motivated reasoning; not ours, but the authors'. Some neglected important factors; others made assumptions that were incorrect; some had basic arithmetic or scientific errors that should have been detected when the results turned out to be so contradictory, but weren't. Rather than exposing our motivated reasoning as scientists, it turned out that those rejecting the science were the ones willing to overlook information in pursuit of their goal.

The authors of those studies hadn't been up front about their motives, but sometimes people can be remarkably honest about this. At a workshop on how climate change affects agriculture in Texas, one farmer came up to me afterward, shaking his head. "Everything you said makes sense, and I'd like to agree with you," he confessed. "But if I agree with you, I have to agree with Al Gore, and I could never do that."

As Peter Boghossian and James Lindsay explain in *How to Have Impossible Conversations*, "think of every conversation as being three conversations at once: about facts, feelings, and identity." I thought I was having a conversation about farming and water; but we were also talking about how we felt about climate change, and about how we saw ourselves in relation to it. "It might appear that the conversation is about facts and ideas," these authors continue, "but you're inevitably having a discussion about morality, and that, in turn, is inevitably a discussion about what it means to be a good or bad person." The farmer had listened to what I'd said and given it a fair shot, and he even agreed with it—logically. But he realized that he'd have to give up his moral judgment to accept this new information. It just wasn't worth it.

HOW FACTS CAN BACKFIRE

In the most extreme cases, when people have already constructed their sense of identity around rejecting so-called liberal solutions to climate change, you can see how bringing up scientific facts can come across as a personal attack on their identity—or an "insult," as my Twitter antagonist termed it. If rejecting climate change is part of what we believe makes us a good person, then we don't interpret arguments to the contrary as "you're wrong." Rather, we hear them saying "you're a bad person." And no one likes to hear that. It tends to make us double down on our denial in a kind of *backfire effect*.

This backfire effect happened in real life, and on cable television, to my friend Anna Jane Joyner. She's a Christian, like me, and a climate activist. Anna Jane's dad, Rick, is a pastor of a conservative megachurch in the southern U.S. He rejects the reality of climate change—and the need for climate action—based on a suite of politically conservative views consistent with those in his tribe: other U.S. white "evangelical" leaders, Republican politicians, right-wing news media pundits, and more.

In 2014 the writers of the Emmy award–winning climate change documentary series *Years of Living Dangerously* figured that the dynamic between Anna Jane and her dad would make for great TV. They brought in actor Ian Somerhalder to interview them about how they'd argued about this over the years—including a six-month period when they weren't speaking to each other. The writers also brought in a climate scientist who's a Christian (me) and a former Republican congressman, Bob Inglis. Bob used to reject climate change himself, but he'd been convinced by his own son that climate change was real and dangerous. Since then, he'd gone on to found republicEn, an organization that advocates for free market solutions to climate change.

Bob and I presented our best arguments. We responded to all of Rick's "But what about . . . ?" and "Gotcha!" questions. We even visited oyster fishermen in Apalachicola Bay, near the extended Joyner family's coastal home, to see firsthand the impacts of a warming ocean on people with no particular political axe to grind. The fishermen were Republicans themselves, just trying to get by. They were worried about how oyster catches were dropping as the oceans warmed, sea level rose, and freshwater inflows declined.

Anna Jane's dad is a smart man. In addition to being the head of a large and successful organization, he is a pilot who understands weather nearly as well as a local meteorologist. And he's also a Dismissive. So what do you think happened as we spoke? Thanks to the social science, you might be able to guess. All of this meant he was *better* at motivated reasoning and *more* likely to be polarized by additional information than the average person, rather than less. And that's exactly what happened.

The more we spoke, the more his rejection hardened. You could see it happen in real time over the course of the episode. He probably felt ganged up on, and I could understand why. He definitely felt that his identity, not his opinions, were being challenged and judged. Unfortunately, the result was to drive Anna Jane's dad even further away, and today his denial is stronger than ever. The same zombie arguments Bob and I responded to back then continue to be hauled out and re-aired at family gatherings, in group text conversations and phone calls.

And it's not entirely his fault, either. It's the way our brains work.

COGNITIVE MISERLINESS AND INFORMATION OVERLOAD

There's so much information available in the world today that there is simply no way our brains can contain everything we need to know.

There's a term for this: nearly all of us are *cognitive misers*. In other words, we look for solutions that take the least thought. And to do that, we often rely on what others think.

I don't know about you, but I don't have enough time and energy to develop and maintain a deep expertise about the nuances of immigration policies, CRISPR gene editing, the latest testimony delivered in front of the U.S. House Judiciary Committee, and what the Canadian prime minister said in his throne speech. I do have a wide array of opinions on such matters, however. I've developed these by listening to my friends and family and colleagues, journalists and podcasters, people and sources whom I trust have spent the time to learn about these issues. And, just like the rest of us, where these trusted figures stand on such issues is often directly linked to their political leanings.

So when it comes to highly polarized topics, as cognitive misers we lean toward accepting the opinions of people we trust, people who share our values. We have an incentive to adopt our tribe's beliefs and opinions, as we are rewarded for doing so. Our reward is both social, through a sense of acceptance and community, and psychological, in that we don't have to research this topic ourselves. Agreeing with others gives us a reassuring sense of certainty, security, and belonging in a world that increasingly seems to be too big and moving too fast. For most of us, the value of belonging far outweighs the value of attaining new information, especially if publicly accepting that information and speaking up might lead to a negative outcome—an argument, the cold shoulder, or even ostracism from your social group. And when we're exposed to information we disagree with—as in one study where researchers looked at how people's attitudes hardened when Democrats on Twitter read a set of conservative tweets and Republicans a set of liberal ones—we tend to double down on our previous beliefs rather than re-examine them.

Neuroscientist Tali Sharot explains in her book *The Influential Mind* that our brains are programmed to "get a kick out of information." But, she goes on, "the tsunami of information we are receiving today can make us even less sensitive to data because we've become accustomed to finding support for absolutely anything we want to believe, with a simple click of the mouse." If we give people new information that contradicts their frame, what they believe, and what their tribe adheres to, their brains just turn off. Even worse, she says, "because we are often exposed to contradicting information and opinions, *this tendency will generate polarization, which will expand with time as people receive more and more information.*"

As consistent as this is with my own experience, I was still utterly horrified as I absorbed Sharot's no-nonsense explanation of how our brains work. Why? Because it suggests that the facts I shared with Anna Jane's dad, and that I share daily on social media, all the information you may share with your family or people you know—even all the climate studies scientists keep publishing with more and more such facts in them—all of these may actually be contributing to the polarization of beliefs about climate change rather than helping to dispel it. *Yikes!*

WHEN FACTS WORK

So does that mean facts are useless? No. Facts are incredibly important because they explain how the world works, whether we like it or not. They are often essential to changing our minds if we are not polarized on the issue. They can even change our minds if we *are* polarized, but only if they can be shared in a way that is able to sidestep the polarization. You saw that already with the story of Tim, the scientist who is also a Christian, like me. Let me give you two more examples.

Kirstin Milks is a high school science teacher in a small, Midwestern university town. Some of her students are the children of university professors; others come from families who self-identify as "rust belt," "hillbilly," or "country." It's like living in—and teaching in—a microcosm of America's political and social divide with respect to issues such as race, climate change, and religious freedom.

She engages her kids on climate change in two effective ways. First, she lets *them* ask the questions. "I ask my students, 'What do you need to know about climate change?' " she says, "and then we frame our learning on climate from those submitted student questions." Previous questions have included "How can our world's adults treat our future this way?" and "What place on Earth is dying the fastest?" and "Do we still have time to change our ways?"

Kirstin tapes these questions up on her cabinet and points to them all semester. The posted questions let her say, "Today we're going to answer this question you asked." She's found this to be very motivating, especially for students who haven't previously seen themselves as knowing about, or caring about, science.

Her second strategy is to give her students the opportunity to carry out the same types of experiments that climate scientists do—by analyzing synthetic ice cores that they build in Pringles cans, or graphing data collected from pond core sediments, or modeling temperature rise with different amounts of greenhouse gases in a computer simulation. Once students see they can follow the lines of reasoning used in the (admittedly more technical) work done by climate scientists, they feel empowered and are much more likely to find climate data trustworthy. "If kids feel like they can do the science themselves," says Kirstin, "it's not a black box anymore."

Extended interaction and engagement with the information, especially when it addresses misconceptions or misunderstandings, can have a tremendous impact: not just on kids, but on their parents, too. Danielle Lawson studies science education. She wondered what impact teaching kids about climate change might have on their parents. So she designed an experiment where middle school students in coastal North Carolina—a relatively conservative part of the state, but one that is exposed to sea level rise, stronger hurricanes, and flooding—were divided into two groups. The teachers for one group integrated climate change instruction into their classes for an extended period of two years; the others didn't.

Before the experiment started, Danielle surveyed the students' parents. Did they think climate change was real? Human-caused? Serious, and we should fix it?—or not? Two years later, Danielle polled the parents again. It turned out that teaching kids about climate change made their parents more concerned about it. Conservative parents changed the most, and daughters were particularly effective at changing their hard-nosed dads' minds. The kids were able to sidestep the polarizations and reach right into their parents' hearts—and minds.

So yes, it's important that we understand and be able to explain that climate change is real, and it's human-caused. But most of the time facts alone are not enough to change people's minds on an issue that touches so profoundly on our identity and morality, that triggers so many of our deepest hopes and fears. So what do we do next?

THE FEAR FACTOR

"Climate change contains none of the clear signals that we require to mobilize our inbuilt sense of threat."

GEORGE MARSHALL, *DON'T EVEN THINK ABOUT IT*

"You do not need to live in fear to get shit done."

TALI HAMILTON, KATHARINE'S STUDENT

As a climate scientist I see things in the data that I find disturbing, concerning, and even frightening. At scientific conferences, we still present esoteric treatises on foraminiferal temperature proxy records and internal variability in multi-model ensembles. But now there are also presentations simply titled "Is the Earth F*cked?," a title so stark that I assume other scientists are feeling this, too.

Satire hits close to home, like the article in *The Onion* describing "a weary group of top climatologists [who] suddenly halted their presentation, let out a long sigh, and stated that the best thing anyone can do at this point is just try to enjoy the next couple decades as much as possible . . . [so] they would be skipping the remainder of the conference to get completely shit-faced at the nearest bar." When I shared this with a roomful of climatologists during a talk to the American Association of Geographers and asked "Who's with me?" 40 percent said they were right behind me; 20 percent said they were already there; and the remainder claimed they just suppress their anxiety.

We shared a fraught laugh and I continued my talk, but it's true that what we see in our scientific work gives us the very opposite of hope. For hundreds of years, we've been living as if there's no tomorrow, running through our resources, putting our entire civilization at risk. And climate scientists are like physicians who have identified a disease that is affecting every member of the human race, including themselves, and no one wants to listen to them.

Our planet has a fever, caused by our lifestyle choices since the dawn of the Industrial Revolution. The news is bad: Antarctic glaciers are accelerating into the ocean, coastal towns are flooding, polar bears are starving, forests are burning, islands are disappearing, and species are going extinct. If we aren't able to change our habits at the fundamental, systemic scale needed, the consequences for humanity will be incalculable.

SHOULD WE BE AFRAID?

Let's stop for a minute and look at this litany of negative news. Is it accurate? Is climate change truly that bad?

I'm a climate scientist and I have to be honest, the answers are: usually, and yes.

We are conducting a truly unprecedented experiment with our planet. And the faster things change, the greater the risks of some really nasty surprises happening. If the Greenland ice sheet destabilizes and totally melts, sea levels will rise by up to seven meters, or twenty-three feet. If enough permafrost in the Arctic thaws, massive releases of heat-trapping methane could put an end to any chance of meeting the Paris Agreement's targets. If ocean circulation slows down too much, from freshwater flowing into the Arctic from melting land-based ice, it will disrupt local climate around the world.

Not only that, but we scientists have a well-known problem: we tend to "err on the side of least drama." If you compare climate predictions with what really happened over the past few decades, you will find that the scientific community gets the changes in global temperature right. But studies have found that it tends to underestimate other observed changes and their resulting impacts. This is especially true of the conclusions of big scientific assessments with hundreds of authors. Why? They all have to agree on their conclusions before they can be published, and scientists are naturally very cautious. We don't like to say something is so, unless we are very, very sure it's going to happen—like 99 percent sure. And we absolutely hate being called alarmist; so that means, these days, we'd prefer to be 99.999 percent sure about anything before we open our mouths.

As I and my coauthor Bob Kopp wrote in "Potential Surprises," the last chapter of the Fourth U.S. National Climate Assessment, "the systematic tendency of climate models to underestimate temperature change during warm paleoclimates suggests that climate models are more likely to underestimate than to overestimate the amount of long-term future change." In other words, chances are that things are going to be worse than scientists say, not better. The planet will survive; the question is, will we?

So, yes: as a scientist it is my considered opinion that there is a very good and entirely objective reason to be afraid. But as a human I believe it's what we do with that fear that makes all the difference.

IS FEAR USEFUL?

When we see people who don't seem to care about climate change, or appear apathetic, we might think, with the best of intentions, "They need to be scared. Let's focus on the worst (true) things we know. Surely this will change people's hearts and minds and spur them to action, right?"

Under some circumstances, it may. First, sharing factually scary information can be an important first step for people who are complacent, who don't think climate change poses a serious threat—or is even real. If you're not worried about climate change, why would you want to fix it? Studies show that pessimistic messages increase risk perception, even among conservatives, and people's belief that they could make a difference. In other words, if they originally thought it was no big deal, learning that it really *was* a big deal made them more concerned and more willing to support action: as it should.

Second, fear works well when coupled with uncertainty to induce *in*-action rather than action. This explains why those who oppose climate action use fear-laden messaging and why they devote such a great deal of effort to trying to cast doubt. "Scientists aren't sure," they say, so why take action on such a questionable issue if the only choices are terrible? "They will destroy the economy and take away our personal freedom to drive cars, eat steaks, and take foreign holidays." Not only that, but we assign a much greater cost to things being taken away from us than we do to obtaining new things. So when we are confronted by something uncertain and fearful, the solutions to which involve personal loss, *nothing* is exactly what we are wired to do.

Third, communicating scary information can also be effective when we're functioning as the "ideal man" envisioned by Plato. If we are making decisions rationally, with emotion following after we've processed the information, then scary facts will cause us to seek a solution rather than to shut down. Often, we like to *think* that's how we think (despite large bodies of psychological research to the contrary). I suspect that's why so much environmental messaging uses a fact- and fear-based approach. As legal scholar Cass Sunstein argues, though, there is a substantial emotional cost to receiving information, which often leads us to metaphorically cover our ears. We'd rather not know about it if we don't think there's anything we can do. And this leads directly to the fourth situation in which fear-based messaging can work: if we *do* know what to do.

THE *UNINHABITABLE EARTH* EFFECT

Fear-based messaging can motivate us very effectively if we know how to turn that fear into tangible action. A practical application of this concept is the following: if negative news about climate change is immediately followed with information explaining how individuals, communities, businesses, or governments can reduce the threat, then this information can empower rather than discourage us. Sometimes we are even able to do this ourselves, internally.

It was fear that motivated New York journalist David Wallace-Wells to write his best-selling 2019 book, *The Uninhabitable Earth*. Perhaps the most exhaustive (and most well-written) compendium of doom-laden climate-change facts outside of the scientific literature, his book doesn't exaggerate or spin the science. It just lays out possible worst-case scenarios in clear, unmistakable, and dire prose, unhampered by scientists' tendency toward hedging our bets and erring on the side of least drama.

After the book came out, David and I discussed it at Climate One, a public forum for conversations about climate change. "The more I learned about the science, the deeper I got into it . . . the more scared I was," he said. But here's where the critical step occurred: rather than curling up in the fetal position, he was motivated to use his skills as a journalist to tell the story, so that other people would have the same reaction.

So, did they? That depends on whether they, like David, already knew what to do about it.

Xiye Bastida is a young climate activist who grew up in the small town of San Pedro Tultepec, about forty-five minutes west of Mexico City. At an early age she began to learn from her father, a member of the indigenous Otomi nation, how to live with the Earth, not from it, and how to protect it in return. But when the lagoon their village depended on for fish was pumped out to supply the big city with water, extended drought and flood soon devastated the local economy and upended her people's way of life. Her family moved to New York City, where they arrived just in time to experience Superstorm Sandy, a record-breaking storm strengthened by unusually warm ocean water and sea level rise. It took a while for Xiye to make the connection to climate change, she said. But when she did, it galvanized her into becoming an activist. She even created a training program for other young people concerned about climate change.

Some years later, in a discussion with environmental journalist Andy Revkin, she said, "I was reading *The Uninhabitable Earth* on a beach and learning that we could be headed towards 7 degrees [Fahrenheit] of warming. It made me sad that my children might never see a beach." It also made her realize, though, that there is a huge difference between a warming of 1, 2, 3, and even 4°C. Even if the world fails to meet the Paris targets, action can still make a big difference, and that's what inspired her.

"I'm not doing this [activism] because I'm sad," Xiye concluded, "but because I'm optimistic about our power to change our course and our ability to come together." She had seen her parents and others responding to environmental issues since she was a child. She was already taking action herself. Recognizing that the worst is not yet inevitable propelled her into further action. That fear worked.

WHEN FEAR DOESN'T WORK

For most of us, though, once fear has called us to attention, we don't know what to do next. And when we're stuck at that point, piling on additional catastrophic stories just fuels our collective sense of helplessness and futility. Take the man Andreas met on the train, for example.

Andreas Karelas is a clean energy expert whose organization, REvolv, helps nonprofit organizations shift to solar energy. In his book *Climate Courage*, which focuses on positive solutions and actions people are taking to solve climate change, Andreas talks about how he saw an elderly man with a big white beard reading *The Uninhabitable Earth* on the train in California.

"I couldn't help but ask, 'What do you think of the book so far?'" Andreas says.

The man replied, "Extraordinary. Even if you are liberal and know about climate change, you realize how uninformed you are."

Curious to hear more, Andreas asked the man how this made him feel.

"Hopeless, because we're not gonna stop it," he said. Then he got off the train.

Many of us might identify with that man. When we start to truly understand the magnitude of the threat climate change poses and the solutions that are needed, our natural response is often fear. Climate change is frequently presented as one of two apocalyptic visions. If climate change continues unchecked, there will be countless millions of refugees, massive droughts and floods, and entire land areas rendered uninhabitable. If we do fix it, many conservatives argue, the economy will be destroyed, and socialism will rule the world. In both visions, no one will be able to eat meat or drive or travel or have children anymore.

Neither sounds great to most people, and the fear and anxiety these possible futures induce are often more conducive to just getting back into bed and pulling the covers over your head than to sustaining long-term action. If we are overexposed to fear-based messages, we can become desensitized. Moreover, there's little evidence that, in and of themselves, they are effective at sustaining long-term action. Rather, fear-based messaging can trigger awareness of our own mortality, invoking our finely tuned package of defenses against the notion of considering our own death—distraction, denial, and rationalization.

MOVING PAST OUR FEARS

To be human is to be a bundle of contradictions—and to have an aversion to anxiety. "We do not accept climate change because we wish to avoid the anxiety it generates," George Marshall writes in his book *Don't Even Think About It: Why Our Brains Are Wired to Ignore Climate Change*. As humans, we are prone to tuning out repeated bad messages if they do not relate directly to our lives, or if we feel like fixing them would be even worse for us than letting them run their course. Our emotional bandwidth is limited.

Digging beyond our emotions into the actual wiring of our brains, hundreds of experiments have shown that humans are literally hardwired to move toward pleasure and away from pain. Translating this into a message directly relevant to climate change, neuroscientist Tali Sharot says, "The human brain is built to associate 'forward' action with a reward, not with avoiding harm, because that is often the most useful response. We're more likely to execute an action when we are antici-

pating something good than when we are anticipating something bad."
And fear also hampers our ability to think creatively. As environmental
engineer and eternal optimist Katie Patrick reminds us: "When the body
releases stress chemicals, the brain shuts down the hippocampus region
and you lose about 30 percent of your brain function, including the creative thinking faculties. Fear and doom shut down your brain capacity
for creative thinking. Vision and optimism super boost it."

Psychologist Renée Lertzman goes further, pointing out that talking
about climate can challenge some of our most deeply held beliefs and
stir up some of our biggest fears—which we typically prefer to avoid
doing. And people who are anxious and alarmed can't remain alarmed
forever. Eventually, we overload and check out. In her TED Talk, *How to
Turn Climate Anxiety into Action*, Renée explains that when we humans
experience stress of any kind, if it becomes more than we can tolerate,
we collapse. One possible outcome of this collapse is depression, despair, and shutting down.

Other possible outcomes of the collapse Renée talks about are anger
and denial. As you can tell, I spend a lot of time trying to understand
denial; understanding the anger, though, comes naturally. I've felt it myself, most memorably after the 2015 Paris climate conference. I'd been
there to support negotiators from poor countries who couldn't afford
to bring their own scientists to the meeting. I spent a lot of time talking with people from those countries, hearing stories of the suffering
they'd already witnessed. Then I went home, back to an environment
where people who claim to share my values think it's okay to close their
eyes and their ears to what's happening in the world. I was so angry that
it took weeks before I could trust myself to have a civil conversation
without feeling like I wanted to scream at them, "How could you be so
selfish? Don't you care?"

Christiana Figueres is a Costa Rican diplomat who shepherded the
climate negotiations that culminated in the Paris Agreement. I was there
for just one event; she lived in this environment for years. In the book
she cowrote with Tom Rivett-Carnac, *The Future We Choose*, she talks
about the grief and pain we experience when we see what is happen-

ing to the world. She offers hard-fought wisdom that parallels Renée's. "Anger that sinks into despair is powerless to make a change," she says. "Anger that evolves into conviction is unstoppable."

Even though they seem radically different, anger and denial can be two sides of the same coin: they are both manifestations of our human response to fear, attempts to control a situation that is wildly beyond our control. But it's what we do with them that matters. When we share scary information about climate change, we're trying to get people to act. Fear does make us sit up and pay attention, at least until our bandwidth is depleted. If people aren't worried about climate change, they should be.

But if we don't immediately connect those fears to people's everyday lived experiences and provide viable and appealing options for dealing with the threat, all too often what happens is exactly the opposite: people disengage or get angry. And if that weren't bad enough, these fear-based information dumps can stimulate another equally two-edged emotion—guilt.

THE GUILT COMPLEX

"No one can unilaterally choose to live in a low carbon economy. The goal is not self-purification but structural change."

LEAH STOKES, *ALL WE CAN SAVE*

"People need energy, and we provide it. We're not the bad guys!"

FOSSIL FUEL EXECUTIVE TO KATHARINE

When we're afraid, and when fear-based messages don't seem to be working well enough (on either ourselves or others), the next step is often guilt. And when we serve up our facts with a side of shame, it gets even worse: because how we judge ourselves and how others judge us is directly linked to our sense of self. Most of us have already decided whether we think something is right or wrong. So when we're shamed for doing what we think is right, that's where it gets ugly.

I experienced this myself, in spades, a few years ago. I was at a brainstorming session in Austin, Texas, with other concerned Christians. We were going around in circles on how to get the climate message out to our community. By mid-afternoon one man, who'd been describing the low-carbon lifestyle he lived in his Catholic community, had had enough. Leaning forward, he fixed us all with a stern gaze.

"All these ideas are very well," he said with emphasis, "but the real problem is *sin*. Every time you turn on your car, *you are sinning*. That is the message we need to share."

My reaction was so visceral, I can still feel its echo today. "Oh really?"

I thought. "So—how did you get here? Don't tell me you walked. Are you saying this meeting is sin?"

And the more I thought about it, the angrier I got. I live in an area with no access to public transportation. So when I take my child to the doctor, as any caring mother would, I'm sinning? When I go to work, as any conscientious teacher would, does he think that's a sin? When I drive to church, as any devout Christian would, that's sinning, too?

He'd meant what he said to be motivating, but it had exactly the opposite effect on me. I felt judged for doing what I thought was good. It made me defensive, and angry.

Why do we hasten to heap guilt on others—and on ourselves? When we do what we perceive to be wrong, we feel bad. When we do what's right and good in our eyes, we feel good. When we judge others and put them down, we feel even better (we are righteous and they aren't!). So when others attempt to impose their value system on us, we understand, fundamentally, that it is about making themselves feel better at our expense. Shaming is a zero-sum game. One person wins only at the expense of another.

You can probably think of a time when you had a similar experience. There you were, doing something you thought was just fine—until someone started yelling at you, judging not only what you were doing but who you were. The unfairness of such experiences can rankle for years and even decades. It's one thing if we knew we were doing something wrong and got found out; it's something entirely different if we were doing the right thing (or thought we were) and were judged and condemned as if we were doing something terrible. No one wants to be called out that way.

WHY PEER PRESSURE WORKS

Don't get me wrong: guilt and embarrassment *can* and often *should be* an appropriate and temporary response to doing something wrong. These emotions remind us of our moral compass and thus serve an important function in human society. As psychologist and marketing expert Robert Cialdini explains, finding out what other people think is one of the most frequent shortcuts we use to determine what is acceptable and what isn't. Awareness of what others think helps us avoid actions that will harm or offend others, such as throwing the contents of our toilet out of the front door as people used to in the Middle Ages, or having a screaming meltdown in the grocery store like a toddler might do when they're out of her favorite cookies. Over time, these habits become ingrained to the point where we all observe them. Since being taught to recycle as a child, I find myself literally unable to throw something out if it's recyclable. I will walk around carrying it until I find a bin to put it in. Not that a single cup matters in the grand scheme of things, but it's the social norm I was raised with and now it's ingrained into my psyche.

As our collective sense of what's considered socially acceptable shifts, it can effect long-term changes in our behavior. For a long time, driving the largest SUV you could find was a status symbol. Now, however, it's more likely to be a speedy electric Tesla than an enormous gas guzzler. In Europe, where the train system is electric, extensive, and fast, there is some evidence that "*flygskam*," or "flight shame," is reducing air travel in Sweden and Germany. People have begun to take trains and explore more local vacation options. And these changes can have a real impact on our personal carbon footprint.

When I first measured mine, though, I was shocked to find that most of my carbon emissions came not from my car or my diet or trips to see family but from flying to scientific meetings, conferences, and climate talks. So I decided to fix that. I set out to be as effective as possible with both my time as well as my carbon. My target was to transition most of my talks to virtual, and fly only if I had enough events in one place that the carbon footprint of traveling to each would be roughly equivalent to driving a reasonable distance from home in my little plug-in hybrid hatchback.

Making the change wasn't easy. But it has significantly paid off in terms of cutting my carbon emissions and increasing my ability to reach people. By the time coronavirus hit in 2020, I was already giving 80 percent of my talks online; and when I do fly, I've assembled as many as two dozen events per trip. When I went to Alaska in fall 2019, for example, I calculated that if I persuaded just eight of the hundreds of people I spoke with there to reduce their own personal carbon footprint by 10 percent, that alone would cover the carbon of my trip. I also offset all my travel with Climate Stewards, a U.K.-based charity whose motto is "reduce everything you can first, then offset the rest." My donations support clean cookstove, agroforestry, tree planting, and ecosystem restoration projects that remove carbon from the atmosphere or prevent it from being emitted in the first place.

Guilt can motivate us to change. Like fear, though, it can shut us down if we carry it with us long-term, or if it's used as a weapon against us. That's what I'd call shame—and I've been on the receiving end of that more than once.

WHY PURITY TESTS ARE UNHELPFUL

Before I travel, I post a list of my scheduled talks online, so as many people as possible will hear about them. A few years ago, when I did this for one of my bundled trips to Alberta, a fellow academic from the U.K. replied immediately, demanding to know why I was not taking the train from Texas. By flying, I had failed his purity test and his response was to shame me.

Now just to be clear, although I know that flying is the biggest part of my personal carbon footprint, I also know that one climate scientist—or even all the climate scientists in the world—never flying again is not going to make a dent in the climate problem. We tried that in 2020. All the coronavirus-related shutdowns together, not just people not flying but all the other reductions from industry and transportation, dropped global carbon emissions by just 7 percent, and even that drop was temporary. We have to make it permanent, and do it every year, to meet the Paris Agreement targets.

But since his response most likely arose from our shared awareness of how bad the situation was, I bit back a sharp retort. Instead, I calculated what it would take for me to take the train from Lubbock, Texas, to Edmonton, Alberta. First, I'd have to drive 6 hours to the nearest train station, in Oklahoma City. Then it would be 57 hours up to New

York City, 12 hours to Toronto and an overnight layover, and another 61 hours to Alberta. By the time I arrived I'd have been travelling for 136 hours—more than five days—and then I'd have to turn around and head back a few days later. I pointed out that if my colleague took a train eastward from the U.K. he'd be in Irkutsk, on the shores of Lake Baikal in Siberia, by the time I got to Alberta. To his credit, he recognized the impossibility of the standard he'd set.

Peer pressure is effective when there is a viable alternative. When there is a bin for the recycling; when we see others using reusable water bottles; when there is a train we can take that will get us where we are going in a reasonable amount of time; when the alternative is relatively (or even more) affordable.

But increasingly, climate guilt is being exploited today less to promote societal good behavior than as a way to shame people, to make ourselves feel better by pointing a judgmental finger at someone else. It's also being used to determine who's in, when it comes to doing the right thing and being a card-carrying member of the "climate action team," and who's out. Shame can be manipulated just as fear can—and perhaps even more damagingly, because it strikes so close to our sense of identity and self-worth. And when there's nothing we can do about it—or at least nothing that seems reasonable, because how do I even get to work without "sinning"?—our reaction is rarely positive. One study found that when people were told to change their behavior, receiving instructions that implied their choices were being judged as inadequate at best, bad at worst, their willingness to take personal actions to reduce their carbon footprint decreased. They were also less likely to support pro-climate political candidates, and even their trust in climate scientists dropped.

The Opower experiment is often cited as a rousing success in the use of peer pressure to effect change. But it, too, ran headlong into this problem. Opower was a software package that allowed utilities to add a small section to each customer's power bill stating how their energy use compared with that of their neighbors. Simply by including this information, the utility companies saw consumer power usage drop by an average of 2 percent, saving a total of over $1 billion in bills over eight years. But here's the catch: a follow-on analysis tracking long-term customer behavior found that "households that were politically conservative and that used more electricity than the norm, did not donate to an environmental organization, and did not pay for renewable energy" *increased* their electricity usage after they got similar information on their bill. If we think we're being shamed into doing something, it makes us feel—or sometimes even do—exactly the opposite.

What social scientist Rebecca Huntley calls the "Puritan ethos of disapproval" that emanates from much environmental messaging can be profoundly counterproductive. But there is an alternative: showing someone that action can make you feel good. Studies have shown how anticipating the pride of making a choice is much more motivating than our guilt at failing to do so. So when people ask me about flying, I don't shame them. Instead, I say there isn't a one-size-fits-all solution and I offer some ideas. For the future, I'm excited about the potential for short-haul electric and long-haul hydrogen- or biofuel-powered flights. But for now, why not "travel thoughtfully"? For some, that might mean not flying. For others, a hybrid approach like mine might be better. Still others might choose to be an advocate: contact an airline and encourage them to accelerate their transition to new low-carbon technology.

In the end, though, my British colleague's comment wasn't really about me. It was about his fear. When we can't control those we really want to— in this case, the airline companies and the fossil fuel corporations and the government and, really, the entire system in which we live—we turn our fear on others and use shame to try to control them instead. We might temporarily feel better, but it just makes things worse long-term.

WHY WE GUILT-TRIP OURSELVES

We even regularly shame ourselves without any help. "I'm a farmer—I need my truck!" said one Dismissive Texan defensively after we'd been talking about how rainfall patterns are shifting. I hadn't mentioned trucks or even fossil fuels or climate change at all: I was just following his lead, talking about how he'd be affected by drought. Even so, his immediate reaction was to defend himself from the guilt his brain, not my words, immediately associated with the issue.

Recently I listened to the results of a survey of suburban women's views on climate change. "Most women describe themselves as pro-environment," said a researcher on the broadcast, "but they immediately follow that with a statement about how they don't do enough, so they're not sure if they really 'count.' They don't recycle enough, they don't drive a small enough car, they aren't vegan and they travel, they say."

Mary Annaïse Heglar works for an environmental nonprofit in New York. In a 2019 essay published in *Vox* titled "I Work in the Environmental Movement. I Don't Care If You Recycle," she describes how people approach her to confess their environmental sins. They don't drive an electric car, they tell her, or they took a vacation this year that required an international flight. And here's what she says:

> I don't blame anyone for wanting absolution. . . . But underneath all that is a far more insidious force. It's the narrative that has both driven and obstructed the climate change conversation for the past several decades. It tells us climate change could have been fixed if we had all just ordered less takeout, used fewer plastic bags, turned off some more lights, planted a few trees, or driven an electric car.
>
> The belief that this enormous, existential problem could have been fixed if all of us had just tweaked our consumptive habits is not only preposterous; it's dangerous. It turns environmentalism into an individual choice defined as sin or virtue, convicting those who don't or can't uphold these ethics.
>
> When people come to me and confess their green sins, as if I were some sort of eco-nun, I want to tell them they are carrying the guilt of the oil and gas industry's crimes. That the weight of our sickly planet is too much for any one person to shoulder. And that that blame paves the road to apathy, which can really seal our doom.

We are all part of the system that depends on the fossil fuels, the deforestation, and the agriculture that are changing our climate. As behavioral scientists Gabrielle Wong-Parodi and Irina Feygina point out, however, this system also provides our safety and security, stability, and meaning. Climate change doesn't just threaten our system, they add. By being a problem that's caused by the same system that keeps us safe, it also poses a threat "to one's sense of personal integrity—a view of oneself as capable, consistent, and adhering to strong moral and ethical principles and values." We want to be good and to be viewed as such, so we respond defensively to any information that may call into question our sense of adequacy or worth, they conclude.

It's no surprise, then, that when it comes to climate change, we feel helpless. We're told that essential aspects of our lives—driving to work, or to the doctor, or feeding our kids, or going on vacation with our family—are bad. But we can't envision how to live otherwise. Or even, how to *exist* otherwise. So when we're shamed, we defend ourselves because we feel, just like I did that day in Austin, that we have no other option. We are just doing our best to get by.

"WE'RE NOT THE BAD GUYS"

Even people who work for fossil fuel companies feel this way. I learned this firsthand the first time I was invited to address the leadership team of an oil and gas company—not an Exxon or Chevron, but one of the big players in oil and gas drilling across the southern U.S.

"Don't worry," said the university alumnus who'd invited me to address their leadership team, "we won't let our chief geologist attend. He's the one who thinks it's all bunk. We want to hear what *you* have to say."

That did not make for an inspiring invitation, so I thought hard about whether or not to accept. If I couldn't identify a core value or belief that I shared with them, I shouldn't be talking to them. This was the advice I gave others—so I'd better do it myself, too. But what did I have in common with oil and gas executives?

Then, like a lightning bolt, it hit me: I am truly grateful for fossil fuels. Without them, I'd be living a life that was far shorter and much more miserable. We don't have to spend our days gathering food for our next meal or worry about it spoiling overnight; we have refrigerators to pre-serve the food we buy at the grocery store. Our means of transportation were once limited to foot, cart, and horse; today we have cars, trains, and planes that move us around the planet at speeds that would shock our ancestors. And thanks to our appliances and our electricity, we no longer have to spend our days on the endless drudgery of menial tasks that oc-cupied so many women a century ago, or wake and go to bed with the Sun. It's not a stretch at all to say that I'm truly grateful for fossil fuels.

When I walked into the executive conference room at the top of the twenty-story headquarters, the atmosphere was strained. A few smiled, but most faces were unwelcoming as everyone took their places.

Nonetheless, I began my talk with those words. And as I said them, I could see everyone around the conference table visibly relax.

"You get it," one man said, disbelievingly, as a smile slowly spread across his face. "People need energy, and we provide it. We're not the bad guys!"

"That's right," I agreed, "and we need energy in the future, too. The question is, how are we going to get it? Because we don't use horses and buggies or party line telephones anymore. And now that we know that fossil fuel use has many serious and even dangerous side effects, the transition is even more urgent. We need to move beyond it as soon as possible. So how do we do that while keeping the lights on and continu-ing to provide local jobs?"

The meeting was scheduled to run for about forty-five minutes, but the discussion was still going strong at two hours. Everyone had questions and wanted to understand: How did we know humans were changing the climate? Where had their geologist gone wrong? And what energy sources might we explore for the future?

MOVING BEYOND FEAR AND GUILT

So what *is* the answer, if it's not piling on the fear with a hefty serving of guilt and shame? Interestingly, neuroscience, psychology, philosophy, and religion all point to the same solution, albeit from different perspectives.

First, we do need to know what's happening. And it *should* make us—not panicked—but seriously concerned. Why would we want to fix something if we don't even know it's broken? "Worry is the wellspring of action," researcher Brandi Morris told me. Her work combines physiology with psychology to study how our brains respond when we learn about climate change. Tony Leiserowitz seconds this. "Fear is not a great predictor of policy support for climate action," he says, "but worry is." And what makes us worried is understanding that yes, climate change is real, it's human-caused, and its risks are serious.

But second, we need to know that we can fix it. There *are* solutions. It also helps to reframe these as "consistent with upholding, rather than destroying, our social system and ensuring its stability and longevity," Wong-Parodi and Feygina find. Being on the same side as the solutions rather than seeing them as in opposition to us is more likely to bring us on board. Going a level deeper into the function of our brains, positive rather than negative reinforcement is key to motivating long-term change. If our brain is hardwired to move forward toward a reward but to freeze in response to fear and anxiety, as Tali Sharot explains, then to spur ourselves and each other to action we must provide a positive incentive to act, not just an apocalypse to avoid.

And third, our burgeoning awareness or knowledge of what's "good" versus "evil" also needs to be able to offer a clear and legitimate pathway to alleviating our guilt. In researching *Don't Even Think About It: How Our Brains Are Wired to Ignore Climate Change*, George Marshall did something very unusual: he decided to tour American megachurches because, he argues, they hold some key lessons on how we think and talk about climate change. Sermons on hellfire and damnation are only effective in spurring action if there's a chance, however slim, of redemption and forgiveness.

WHAT FAITH CAN TEACH US

As a Christian myself (though not one who attends a megachurch), I find it particularly striking how the Bible talks about fear and action. In the apostle Paul's letter to Timothy, he states simply that God has not given us a spirit of fear. So if we feel fear and respond to climate change out of fear—fear of either the solutions or the overwhelming nature of the impacts—that fear is not coming from God. Instead, Paul continues, God has given us a spirit of power, which enables us to act, instead of being frozen or paralyzed. We have been given a spirit of love, to have compassion for others, which means caring for others, putting their needs first as we act. And finally, we have been gifted with a sound mind that we can use to make good decisions based on facts and data that God has given us and made evident to his creation.

So how do we move beyond fear or shame? By acting from love, I believe. Love starts with speaking truth: making people fully aware of the risks and the choices they face in a manner that is relevant and practical to them. But it also offers compassion, understanding, and acceptance: the opposite of guilt and shame. Love bolsters our courage, too; what will we not do for those and that we love? And finally, it opens the door to that most ephemeral and sought after of emotions, hope.

Hannah Malcolm is a theologian who grew up listening to her grandfather, pioneering climate scientist Sir John Houghton, talk about the urgent problems facing humanity. "Whole countries will be underwater in fifty years if we don't do something now!" she recalls him saying. She sees the echo of modern scientists' warnings in the apocalyptic language of biblical prophets, warning of catastrophe if the status quo continues. But she points to a key next step: "The words of the prophets—living and dead—can help us learn to talk about our apocalyptic fears. They teach us to be honest about the realities of sin, greed, and grief. They call for radical, upside-down changes, not small adjustments to existing systems. *And they teach us how to be absurdly hopeful, painting visions of peaceful futures when that seems impossible.*"

SECTION 3:

THE THREAT MULTIPLIER

A FARAWAY THREAT

"We are navigating recklessly towards our future using conceptions of time as primitive as a world map from the fourteenth century, where dragons lurked around the edges of a flat earth."

MARCIA BJORNERUD, TIMEFULNESS

"We've looked at what our own data is telling us, and it's very clear. It's getting warmer and we have to prepare or we'll be hung out to dry."

MAN AT A WATER PLANNING MEETING IN TEXAS

Over scrambled eggs at a diner in Salt Lake City, I was being treated to some of the wildest polar bear stories I'd ever heard. I knew they were all true, though. Steve Amstrup is the chief scientist for Polar Bears International, and his work tagging and examining hundreds of individual bears and tallying their numbers has led to polar bears being listed under the U.S. Endangered Species Act. We both happened to be in Utah to give talks later that day, so we'd seized the chance to meet up for breakfast.

I jokingly asked Steve how many bears he'd given the kiss of life. Instead of laughing, he did some mental math before replying, "As many as a dozen." And then he told me about the trip his team takes every fall to Churchill, Manitoba, to observe the bears before they head out on the ice for the winter.

"Why not come see the bears for yourself?" he asked.

I wanted to go—who wouldn't? But I already had a hectic schedule planned for the fall, including the Paris climate conference in December, and my work focuses on how climate change affects people, real humans, in the here and now. Not only that, but I believed that when polar bears are used as the symbol for climate change, it does the rest of us a disservice by making the issue seem remote and distant. Melting glaciers and starving bears are real, and serious. But genuine concern about climate change, concern that motivates long-term action, usually has to be based on something closer to home.

My reluctance must have shown on my face because Steve then said something that completely changed my perspective. "We care about the polar bears because they're showing us what's going to happen to us," he said. "If we don't heed their warning, we're next."

That's when I decided to go after all.

WHAT POLAR BEARS CAN TEACH US

The life of a polar bear revolves around sea ice. It's where they feed in the winter on seals, their preferred prey. But today, Arctic sea ice is in a kind of death spiral. As the top of the world warms, its ice cap thaws, exposing the ocean beneath it. That dark water absorbs more of the sun's energy than the reflective white ice—so the Arctic heats up even more, triggering a cycle that is causing it to warm twice as fast as the rest of the planet. Arctic sea ice is declining by an area the size of Ireland, on average, every year. In 2020 it hit the second-lowest extent on record, almost half what it was when the satellite record began in 1979. Submarine and satellite measurements show that the average thickness of the ice has also declined by nearly half since 1958.

The bears' feeding ground is literally melting. As sea ice disappears earlier every spring and forms later each fall, more polar bears are spending more time on shore, fasting. But the prey they catch on land isn't a viable substitute for what they catch on the ice. That's why polar bears are one of the first and most visible species to suffer the effects of a warming climate. When I went to the Arctic with Steve's team in 2015, I saw this with my own eyes.

Historically, the ice on Hudson Bay refreezes in early November. But when we arrived in Churchill after a nearly two-thousand-mile trip, there wasn't a piece of ice in sight, just plenty of ravenous bears. My son, then eight years old, had come along. Our second night there was Halloween, and he was wide-eyed at the sight of grown-ups patrolling to keep trick-or-treaters safe from bears that often stray too close to town. During our first morning out on the tundra, he shook me awake as the sun appeared over the horizon. "Look outside!" he said, pointing. "This bear has been waking me up all night, standing up and peering in the window at us." And, sure enough, there was a giant bear right outside the window of the tundra buggy: curious, bored—and hungry.

Many consequences of climate change are far more subtle than a famished bear inches from a third-grader, but they are no less proximate and life-threatening. And they impact our food sources, not just the bears'. In her TED talk, *How Climate Change Could Make Our Food Less Nutritious*, public health researcher Kristie Ebi explains how higher carbon dioxide levels make plants such as rice, wheat, and other crops grow faster; but as they do, their protein and nutrient levels decrease. Then there's the impact of warming temperatures on pests and yields; from 1981 to 2002, for example, it's estimated that climate change was responsible for an average annual loss of $5 billion worth of wheat, maize, and barley around the world. These crop losses often happen in poor countries where people already live on a few dollars a day. When the price of food doubles, families go hungry.

Steve was right: what's happening to the bears is happening to people, too. Yet all too often, we seem to be even less conscious of it than the bears.

PSYCHOLOGICAL DISTANCE

The idea that we're invulnerable to anything the planet might throw at us isn't unique to climate change. In Lubbock, Texas, where I live, no one doubts the reality of tornadoes. Yet as tornado warnings went out on May 11, 1970, veteran west Texas broadcaster Bob Nash dismissed them, saying, "You have less chance of being hit by a tornado than being trampled by a dinosaur." Within hours, up to a quarter of the buildings in the city had been damaged or destroyed, more than thirty people had lost their lives, and hundreds had been injured. To this day, the 1970 Lubbock tornado remains one of the strongest tornadoes to hit the business district of any American city.

This very human tendency to ignore certain types of threat is called *psychological distance*. It's part of a theory that posits that the further away something is from us—in time, or physical distance, or social relevance—the more abstract and unimportant we will consider it to be. In contrast, the closer something is to us, the more concrete and more relevant it appears.

This explains how effective it was, for example, for Senator Jim Inhofe, an Oklahoma Republican and longtime climate Dismissive, to hold up a snowball in the U.S. Senate in 2015 while claiming that global warming wasn't real. A real snowball dripping onto the carpet of the Senate made for a much more salient, physical example to many people than decades of temperature data showing that Washington, D.C., the U.S., and the world are warming.

WHY CLIMATE CHANGE SEEMS SO FAR AWAY

More information about why climate is changing, or even information about its impacts on polar bears, may satisfy our curiosity. But the concept of psychological distance explains why it doesn't necessarily make us more concerned about climate change or more willing to support or engage in climate action. Climate change falls prey to nearly all the types of distancing explained by this theory.

First, it's abstract rather than concrete. Unlike air pollution, climate change is caused by invisible heat-trapping gases that we can't see, feel, or smell. Compounding the problem is that it's typically represented by global temperature. Getting a handle on that requires adding up daily records from thousands of weather stations around the world for at least a few decades: a nebulous concept, as compared to the weather here and now.

Then there's the issue of actual distance, in space and time. People often think of climate change as something that happens to people and places far away: to Steve's polar bears or to people who live on low-lying islands in the South Pacific with unfamiliar names, like Tuvalu or Nauru—but not to them. They also think about it as something that will affect their children or grandchildren sometime in the future, but not them, today.

And finally, there's the issue of social relevance. Global warming is often perceived as a niche issue. We think it's something that matters to people who proudly own the label of "environmentalist" or "tree hugger" or "save-the-whales campaigner," people who vote for Bernie Sanders or their country's Green Party. But if we're not people who would describe ourselves this way, then all too often we don't think it matters to us.

The Yale Program on Climate Communication tracks public opinion throughout the U.S. and Canada across more than two dozen questions related to climate change. You can still see the effects of psychological distance in their regional polling data. As of 2020, over 70 percent of people in the U.S. agree that global warming is happening and that it will harm plants and animals (things that are not as relevant to us as our own lives) and future generations (people in the future, not now). Sixty-five percent of people agree global warming will harm people in developing countries (who live far away) and 61 percent even say that it will harm people in the U.S. (who are not them). But when the Yale researchers ask "Do you think climate change will harm you personally?" the percentage drops precipitously, to a mere 43 percent. Somehow, the majority of us imagine that climate change will affect the world we live in, people far away, even our grandchildren and our neighbors, but not us.

HOW TO FLIP OUR PERSPECTIVE

"When it comes to rare probabilities, our mind is not designed to get things quite right. For the residents of a planet that may be exposed to events no one has yet experienced, this is not good news," writes psychologist Daniel Kahneman in his book *Thinking, Fast and Slow*. In other words, when we live in rare times, as now, the way we think puts us at grave risk of underestimating the harm we might face—which in turn increases our risk.

Psychological distance is a more widespread challenge than whether or not we accept the science of climate change. Many of us who acknowledge that global warming is happening still see it as just one more item on our overflowing list of priorities. News headlines are full of urgent problems: a global pandemic and looming economic crisis; refugees and immigration; water, energy, and our finite resources. As individuals our daily attention goes to our health, our safety, our jobs, and our families.

This might sound daunting, but Chris Chu, a communications researcher from my own university, Texas Tech, has flipped the issue around. Rather than looking at why we distance ourselves from climate change, he studies how we can stop doing it. His work confirms that when we tell people about what's happening on the other side of the world—telling Americans about the impact of climate change on Indonesia, for example—they see it as distant and less relevant to their lives. Our strong frames of political polarization continue to impose their overwhelming filters on the information we receive and dominate our opinions on climate change.

But when we talk about what's happening here, locally, in ways that matter to us, all of a sudden our perspective shifts. For example, when we understand how local sea level rise affects us, we're more willing to cut our carbon footprint. Not only do we recognize climate change's relevance to our lives, but our political polarization decreases. In other words, we are more likely to agree that climate change matters when its impacts are presented to us as here and now.

When we reduce our psychological distance, what we have in common starts to become more important and more relevant than the political ideology that divides us. This is a key insight that signals a way we can alter our approach to talking about climate change and begin advocating for climate action more successfully than with the strategies we've used in the past.

TALKING CLIMATE CLOSE TO HOME

I've seen firsthand the amazing depolarizing power of sharing local climate impacts. A few years ago, I was brought in to help develop climate projections as part of a long-term plan for a central Texas water district. The district's leadership team was skeptical that climate information was needed, so I was asked to give a preliminary talk before they decided on the proposed work. The introduction they gave me was a surprise: it had little to do with me or the proposed work, other than mentioning my name. Instead, it was all about my university: what a good school it was, how successful our students were, and how well our football and basketball teams were doing. As I looked around the room at all the smiling faces, the penny dropped. Nearly everyone there was an alumnus. So even though some were still somewhat suspicious of me due to my title of climate scientist, I was considered to be more trustworthy than a scientist from a rival institution because I was a member of their "tribe." We supported the same teams. We could sidestep the polarization, just like those kids from North Carolina did with their parents. It sealed the deal.

I didn't take this accord for granted. Water districts are very knowledgeable about data; so throughout the project, I made sure that the water gauges and the weather stations our team were using were ones the district knew and used, too. Once we understood the historical data, we calculated the impact of a 2°C and a 4°C warmer world in terms of measures relevant to them: water supply, consumer demand, and return flows. Our results highlighted how successful the district had been at conservation efforts in the past. But they also showed that in a 4°C warmer world it would be difficult, if not impossible, to continue to provide the same amount and quality of water as they do now. In a 2°C warmer world, on the other hand, responses were costly, but within their range of viable long-term options. They could increase conservation and add some new sources to stabilize their supply during longer, drier periods.

When the project engineers presented the final results to the board, there was an animated discussion but argument was minimal. Everyone recognized the names of the weather stations and the locations of the gauges used in the analysis. All the graphs showed historical data and future projections together so they could see how they compared. Clearly, past action had paid off, and more such action was needed; everyone agreed.

Then possible responses were brought forward, including the question of how to communicate these findings to the people in the district. At that point, there was a long silence. One of the board members finally voiced what they were all thinking.

"We're in a pretty conservative area," he said, "and this here what we're talking about is global warming. How are folks going to take it when we start telling them this?"

But the oldest, most senior board member was having none of it.

"I know exactly what you're talking about," he said. "My granddaughter comes home from school telling me things she's learned, like the polar bears are endangered because the ice is melting. I try to tell her that's not true, but she won't listen to me."

Heads around the table wagged sadly: the gullibility of the younger generation and the propensity of the education system to brainwash its students was evidently not a new topic.

"But this here is different," the older man said emphatically. "In this case we've looked at what our own data is telling us, and it's very clear. The data's the data. It's getting warmer and we have to prepare or else we'll be hung out to dry."

He pounded the table in emphasis and repeated again,

"The data's the *data*."

And that was the final word.

CLIMATE CHANGE IS THE CURVE IN THE ROAD

Climate change affects us in many different ways, depending on where we live and what we care about. But the fundamental cause of all our problems is the fact that we—both humans and polar bears—are wholly unaccustomed to, and unprepared for, climate changing this fast. Here's how I picture it.

Where I live in west Texas, it's so flat that many of the roads are dead straight. They're so straight, in fact, that you could drive along looking in your rearview mirror the whole time. Why? Because driving down a straight road, where you've been in the past is a reliable guide for where you'll be in the future. But what happens if you're driving down the road looking in the rearview mirror and you hit a curve? Let's just say, you'll end up somewhere you didn't plan to be, and it won't be good news. And that's exactly what's happening to us humans as we drive down the climate road today.

Modern civilization is built on the assumption that we have a straight road and a stable climate, and that therefore the conditions we've experienced in the past, that we can still see in our rearview mirror, are accurate predictors of the future. Based on what we see in the mirror, we've parceled out our land, developed complex agricultural systems, constructed trillions of dollars' worth of infrastructure, and allocated our water resources.

Over the past few thousand years there have been small, regional-scale climate variations or wiggles in the road. Some of these had important local and even regional consequences. Sometimes they presented new opportunities, such as the Medieval Warm Period I talked about in Chapter 4. Other times, shifts in rainfall and temperature contributed to the decline of civilizations such as the Mayans in Central America, the Anasazi in the Southwest U.S., and the Bronze Age Indus Valley civilization in modern-day Pakistan. But for the most part we've been successful at staying on the road because it was largely straight. *Was.*

Today, the Earth has warmed by over 1°C (nearly 2°F) in the past hundred and fifty years. This rate is over ten times faster than when the Earth emerged from the last ice age. The planet is going to survive—it has, after all, been warmer before on geologic timescales. It is our human systems that are at risk, our cities and economies and buildings and food systems and, at the end of it all, our civilization.

That's why it's past time to take our eyes off the rearview mirror and take a long, hard look at what our future has in store. Right now, it's as if we're all piled into a giant bus barreling down that road, heading for a very big curve. Our wheels are on the rumble strip and they're sounding the alarm. And as Steve the polar bear scientist said: "If we don't heed the bears' warning, we're next."

HERE AND NOW

"Ours is the first generation to deeply understand the damage we have been doing to our planetary household, and probably the last generation with the chance to do something transformative about it."

KATE RAWORTH, *DOUGHNUT ECONOMICS*

"I've lived here for thirty years, and the weather is just getting weirder."

MAN AT CHURCH TO KATHARINE

Back in 2013, the Weather Channel had a competition for the wildest weather city in America. Lubbock, Texas, won the whole thing, besting Fargo, North Dakota; Fairbanks, Alaska; and Caribou, Maine.

If you know weather, that's no surprise. When you look at the number of climate and weather disasters that have caused at least a billion dollars of damage since the 1980s in every state across the U.S., Texas has had the most of them. As of 2020, it's experienced 124 of these events in forty years. That's an average of 3.1 per year. If you think of extreme weather as an unlucky chance event, like a roll of the dice, then Texas is already playing with a weighted pair.

Just before Christmas the other year, I was invited to speak to a women's group near my home. It was a popular group: seasonal baking overflowed across several tables in the dining room; every other square inch was covered in decorative wooden nutcrackers; and all the seats were filled. There wasn't any room for a projector or slides. It was just me, perched beside the fireplace surrounded by women, a very large Christmas tree, and the nutcrackers.

I usually rely on visual aids in my talks, whether a diagram showing how the Earth's temperature and the Sun's energy are going in different directions, an image of the aftermath of Hurricane Harvey, or a photo of a woman in Bangladesh holding up the solar panel that helps her run a small business and send her children to school. Because I couldn't use any this time, I decided to use the images people already had in their minds. Weather can create some vivid ones. I've been here in Lubbock more than ten years myself, and I've already got over a dozen stories of floods and droughts, ice storms and thunder-snows, hail, mud-rain, and heat waves.

―――――

I started by telling my own worst weather story. In October 2011, I had to catch a morning flight to a research meeting in Denver. It was still dark when I was woken by a call from United Airlines. The direct flight from Lubbock to Denver had been canceled. "But if you get to the airport by 5:55 a.m.," she added, "you can still connect to Denver through Houston." It was already 5 a.m.—so I threw on my clothes, grabbed my computer bag, hurled myself into the car, raced to the airport, opened my window to grab a parking ticket . . . and forgot to close it back up again.

I made the flight. But later that day, a historical, record-setting, national-news-making haboob—a vertical wall of dust that you can see coming from miles away, capable of turning daytime into night and blanketing the entire city in a thick cover of dirt—rolled over Lubbock. It went right through my car, sitting in the exposed airport parking lot. The open window happened to be facing due west, precisely the direction the haboob blew in from.

When I got back, I could not find that parking ticket anywhere. In fact, I couldn't find anything that had been in my car when I left it. The entire interior was coated in a layer of red dust. For the rest of its life, the air vents in that car squeaked as if there were a cricket stuck in them. I'll never forget that experience.

CLIMATE CHANGE IS LOADING
THE WEATHER DICE AGAINST US

I then asked each woman to share her most memorable weather story. Everyone had them, and many found it hard to choose which one to share. One woman spoke of a dust storm during the 1950s so thick it turned the midday sky black; another described the infamous Lubbock tornado that destroyed downtown and twisted the tallest building right around, as it remains to this day; others shared stories about all the droughts, floods, and hurricanes they had lived through. The images their stories evoked were so vivid that I could easily picture them.

How does this connect to climate change? Here in Texas—and in fact everywhere—climate change is supersizing many of our weather events, making them stronger, longer, and more damaging. It's loading our weather dice against us. Two or three decades ago, we'd have been hard-pressed to identify a way that climate change was impacting the places where we lived. Today, though, no matter where we live, we've personally experienced the effects of a warming planet weighting our weather dice.

Heat waves are stronger, and droughts are longer. In the summer of 2019 alone, there were over four hundred all-time-high temperature records broken, in twenty-nine countries, across the Northern Hemisphere. Climate change also amplifies wildfires from Australia to Alaska, making seasons longer and causing fires to burn a greater area. It's supercharging our tropical cyclones and hurricanes so that they're bigger, slower, and stronger. When California is breaking a new record every year for the largest wildfire ever, when a Siberian heatwave tops 38°C, or 100°F, and when hurricanes are dumping more than 1.3 meters or fifty inches of rain on the Gulf Coast? *That's* climate change.

So I told the women in Lubbock, that day: if you live in the Texas cities along the Gulf Coast, you care about rising seas and stronger hurricanes. And it's not just us. Around the world, there are hundreds of millions of people who live in places threatened by rising sea levels and stronger tropical storms such as hurricanes, typhoons, and cyclones. If you live in the High Plains, in this agriculture- and resource-intensive economy we've built in a semidesert region, you care about stronger droughts and more extreme summer temperatures. Hundreds more millions live in places that are already water-scarce, and it's just going to get worse for them, too. If you care about any of those things, then you already care about climate change. Heads nodded: it fit with what they'd seen for themselves, and it made sense.

Just a few weeks before, I'd been waiting in line after church to pick up my son from Sunday School when the man behind me asked me, out of the blue, "Do you think our weather is getting weirder?"

"Yes," I confirmed. "when you look at the data, it really is."

"I knew it," he said triumphantly. "I've lived here for thirty years, and I can see it."

The women at the Christmas party thought so, too. Even people who aren't sure about global warming are pretty convinced that we're witnessing an increasing amount of global *weirding*.

THE COASTS ARE OUR CANARIES

Climate change is affecting people where I live in Texas. It affects you, wherever you live, anywhere in the world. But those on the coastlines are the canaries in the coal mine: they are the ones to whom the impacts of climate change are already most obvious. Global sea level has already risen by almost twenty-five centimeters (ten inches) since 1880. It's rising as land-based ice from glaciers and ice sheets melts, and because the oceans are heating up and warmer water takes up more space. The trend is accelerating: sea level is rising thirty percent faster now as when the satellite record first began over thirty years ago. Like the polar bears, what's happening today to those who live along the coasts is what will happen to the rest of us tomorrow. That's why it's past time to not only worry about climate change, but to start doing something about it.

Princeville, North Carolina, is remarkable for being the first town incorporated by African Americans in the United States. It stands on the low-lying swampland the white settlers didn't want. In the 1900s, a series of levees were built to protect it from frequent floods, but today those levees are failing. Back-to-back hurricanes put most of the town underwater in 1999, and again in 2016. "If it comes again," says resident Marvin Dancy, who's already rebuilt his home twice, "I don't think I am coming back." Many from Houston, Texas, might agree with him. Some parts of the city have experienced three five-hundred-year flood events in as many years, as urban development, sinking land, and heavier rainfall due to climate change have loaded the dice for them as well.

Forty percent of people in the U.S. live in coastal communities. Around the world, it's estimated that 700 million live in the low-elevation coastal zone; it's home to most of the world's megacities as well. Low-lying islands such as the Maldives in the Indian Ocean or Kiribati and Tuvalu in the South Pacific are particularly at risk. Perched just a few feet above sea level, with the maximum height of their island nations measured in centimeters or inches rather than meters or yards, it doesn't take much to overwhelm them.

Simon Donner is a Canadian climate scientist who's originally from Ontario, like me. When not biking to work at the University of British Columbia, he's diving and studying coral reefs in the South Pacific. Stronger and more damaging ocean heat waves are increasing dangerous coral reef bleaching events. If ocean water becomes too warm, corals expel their algae, turning white. As the algae supply much of the coral's energy, this threatens the survival of the reef. So Simon and his research group are developing a Pacific-wide database of past coral bleaching events so scientists can understand their impacts and how to best protect the coral in the future.

In 2005, Simon arrived in the atoll nation of Kiribati to survey its reefs—just as it was hit by its worst storm on record. With a maximum elevation of just three meters, or ten feet, Kiribati is naturally vulnerable to huge storm surges that regularly occur during El Niño events. This time, though, high winds and high tides, coming from the worst possible direction, sent water pouring over the causeway that connects the islands. Water smashed through seawalls protecting landfills, flooded homes and the hospital, and sent pigs and other livestock squealing through the flooded streets. What struck him most, though, Simon told me, was not the storm. It was the way people reacted, stoically going out to help one another patch roofs and clean debris from the road, doing what was needed so life could go on.

The day after the storm, his friend Taratau Kirata took Simon out to his family's land. The storm had damaged the coral rock wall they'd built, broken the legs off their bwia, an open-air sitting and sleeping platform, and eroded away most of the sand beneath, leaving a pit full of washed-up branches. But Taratau wasn't upset. He just said, "We need to build a new one."

"That's the i-Kiribati people I've come to know," Simon told me, "strong and resilient, but facing the fight of their lives. The pictures we see of coastal flooding are real, but they don't tell the whole story because now they've got sea level rise and the impacts of a warming ocean added to the mix." Simon still studies coral reefs—but since then, he's also been working with local government and communities in Kiribati, to help them understand and prepare for the effects of climate change on their islands and peoples.

THE ARCTIC HAS CANARIES, TOO

Another four million people make their home at the top of the world, in the Arctic. Many of them are finding the ground shifting beneath their feet. Permafrost is constantly frozen ground that doesn't thaw, even in the summer; hence its name. But temperatures across the Arctic are increasing twice as fast as the global average, so a lot of that permafrost is now starting to thaw. Thawing permafrost turns into quagmire, unstable ground for anything built on it. This leads to warped asphalt, cracked building foundations, and, ironically, imperiled pipelines. It's estimated that Russia is already spending about $2 billion a year to shore up oil and gas pipelines on thawing permafrost in Siberia.

In 2003, the U.S. Army Corps of Engineers found that most of the more than two hundred Native American villages in Alaska were already at imminent risk. Many remote communities in the far north are accessed solely by "winter roads" (such as those made famous in the

History Channel TV series *Ice Road Truckers*) or rail lines. Villagers depend on these to stockpile basic necessities such as food, fuel, and building materials. As the Arctic warms, the winter road season is getting shorter, isolating many of these communities and cutting off key supply chains and transportation routes. In Canada, the cost of replacing these with all-season roads is estimated at $373,000 Canadian per kilometer of road or $600,000 U.S. per mile.

Terry Chapin is an ecologist who's been tracking the impacts of a warming planet on Alaska for over fifty years. He has a bushy gray beard and, like many Alaskans, favors plaid shirts, fleece, and jeans. A former president of the Ecological Society of America, he created the concept of "earth stewardship" and, like most scientists, he's worried about the future and wants everyone to understand why. So in 2019, he invited me to Alaska to share what I knew with community and faith groups throughout the state—and he wanted to show me what he and his colleagues were seeing, too.

TRACKING THE TICKING METHANE BOMB

I first arrived in Fairbanks, where we piled into a well-worn university van along with a group of high school students. With Terry at the wheel, we drove north over the roller coaster roads to the U.S. Army Corps of Engineers' Permafrost Tunnel Research Facility. Chris Hiemstra, one of the scientists who works on this underground passageway, cheerfully handed us hard hats and instructed us to bundle ourselves into extra jackets before opening the door to a whole new world: a tunnel of frozen earth stretching deep into the darkness, with veins of ice running through it. Some veins were just a few fingers wide; others stretched all the way up the wall and overhead.

As permafrost thaws, so, too, do these giant veins of ice. That leaves huge cavities under the ground: without a proper foundation, roads can quickly drop a few feet or more. The permafrost under the roads in many places has thawed and refrozen many times. That's why the roads surrounding Fairbanks were the bumpiest I've ever seen. Our hotel was full of road crews who'd been brought in to fix them. But it was often a futile task, considering that the next thaw would just crack them even further.

Entering the slanting permafrost tunnel, it was hard to comprehend the wonder of everything I was seeing. "This is the closest I'll ever come to time travel," I thought, as the students and I traversed the metal walkway through the tunnel. The tunnel's portal had been cut through a layer of permafrost from a few thousand years ago, but by the end of the tunnel we had traveled back some forty thousand years.

As we walked, I noticed bones sticking out of the rough-hewn layers of ice and soil that made up the walls, from animals that had roamed the tundra millennia ago. "What's that one?" I asked Chris, pointing to a jagged piece of bone. Casually, he replied, "Oh, that's probably the leg bone of a mammoth." Hanging from the roof were the roots of plants that were thousands of years old. They were still intact and, for Terry, completely recognizable. "Shrub willow," he said, touching a set of frozen dirt-covered roots hanging down from the ceiling.

Farther down, along a shelf of ice that stuck out into the tunnel at eye level, Chris reached up and gently lifted a few perfectly intact and preserved leaves off the ice. They had dropped out of the permafrost in the last few days, he said. He put them in my hand; they looked like they had just fallen off the tree and shriveled up that week. They were still tinged with green, and you could see the veins. Willow again, said Terry. Yup, said Chris, probably about twenty to thirty thousand years old at this depth.

The entire tunnel smelled of very old cheese. When I was leaving, I finally asked why. Chris said they weren't totally sure, but it was probably the smell of thousands-of-years-old organic material, including poop, thawing.

The plants and animal remains and, yes, the poop are why thawing permafrost matters to us all. As the organic material from plants and animals from so long ago begins to thaw and decompose, it produces methane, a gas that is thirty-five times more effective at trapping heat than carbon dioxide. And scientists are measuring it coming out of the thawing ground in the Arctic faster and faster.

After the tunnel, Terry loaded us all back in the van, and we drove out to Katey Walter Anthony's study site at a lake outside Fairbanks. Katey is a biogeochemist who's trying to get a handle on exactly how much methane is leaking into the atmosphere as the permafrost thaws. She's built a methane collection system from garbage she originally collected while working in Siberia: a thick tarpaulin with a hole in the middle that opens into an empty two-liter soda bottle, a tap on the end of the bottle that can be opened or closed, and all of this held together with some elastic bands and a stick. She submerges the tarp in the water while a student stirs up the mud at the bottom of the lake with the stick. This releases the methane that's been seeping up from the thawing ground under the lake. As it bubbles up through the water, it's trapped by the tarp and funneled into the empty soda bottle. When the bottle's about two-thirds full, she opens the tap and lights a match. Boom!

Every year, it seems, the numbers tick up. If human emissions continue unchecked, by the end of this century, scientists like Katey estimate that most of the world's surface permafrost could thaw. This will release even more heat-trapping gases into the atmosphere that amplify the direct impacts of human-induced warming. As Katey says, these emissions are like a headwind our mitigation efforts are fighting against. And that affects us all—not just those of us who live in the Arctic.

INTRODUCING SOLASTALGIA

There's a name for the mental or existential distress of our environment being changed in unwelcome ways. It's *solastalgia*, and I heard it again and again as I traveled through Alaska, from people who could see their home changing literally before their eyes.

After Fairbanks, I went on to Anchorage. As I walked out of the airport, I could see someone at the bottom of the elevators jumping up and down holding a big welcome sign. It was Scott, who for years now had been providing patient and kind explanations to people on my Facebook page who were trying to pick a fight over the science.

Scott is an engineer who has lived in Alaska all his life. He told us how he grew up hiking its glaciers with his parents, as he drove caribou ecologist Tim Fullman and me up to the Byron Glacier. Its "observing station" is now far from the glacier's edge. To catch a glimpse of the glacier through the blinding rain, we had to scramble across the slippery jumbled black rocks and ledges. These are uncovered as the glacier retreats faster and faster up the mountain, leaving the valley free of permanent ice and snow. Scott said he feels as if he's missing a friend; what he loves is disappearing.

A few days later, in the coastal city of Juneau, Linda took me up to the highest point above the city. She's a retiree who's become a strong climate advocate in her community. She pointed across the channel to the mountains she and I could see on the other side. "Those are the Tongass peaks," she said, blinking away a tear. "I've lived in Juneau for forty years. And every day of those forty years, whenever I looked out across the water, I could see the snowy peaks. But three years ago, that snow melted. Land was revealed that no human eye had ever seen, not even the Tlingit people whose descendants still live here today." The Tongass Mountains are home to one of the few remaining old-growth temperate rainforests on the planet. The mountains were already threatened by logging; now, the loss of their summer snowpack puts their water supply at risk.

WHY SEEING IS BELIEVING

Climate change isn't a future issue. It is here and now, for all of us. When we're able to see its impacts with our own eyes, and understand what we're looking at, this experience can breach many of the emotional and political frames we've built up in our minds.

My last talk in Anchorage was at a local evangelical church. There were some other Texans there that day: David Schechter, host of a television show called *Verify* that gets to the bottom of issues people aren't sure about; his video crew; and Justin, a roofer from Dallas. The point of the show was to see if they could take someone who was extremely doubtful about climate change, although not a Dismissive, and change his mind.

A month before, David had taken Justin on a road trip around Texas. They sat down with me, with my colleague Andy Dessler at Texas A&M University, who's an expert on climate modeling and policy, and with another colleague, Jay Banner from the University of Texas at Austin, who studies past climate using data from stalactites and stalagmites in caves. Justin and David also spoke with one of the few U.S. scientists with expertise in climate who dismiss the science and downplay the impacts—but that had to be by video, because we don't have any such scientists in Texas.

After talking to the scientists and learning about the science and seeing how climate change was affecting Texas, Justin moved from being doubtful to being cautious—clearly, there was a lot more to this than he'd thought. But it was still hard to overcome a lifetime of thinking that this was just a liberal hoax. So David took him to Alaska, where they could see the thawing permafrost and receding glaciers with their own eyes. They hiked out to one glacier with Brian Brettschneider, a climatologist who told Justin all about the crazy high temperature records that had been broken that very year. He explained how sea ice around the state melted completely for the first time in recorded history and wildfires had devastated the forests around Anchorage.

Then Justin attended my talk at the church, where I focused on how climate change was affecting real people, today. What we believe as Christians, about being good stewards or caretakers of the planet and caring for those less fortunate than us, is exactly why climate change matters to us.

I could see his mind connecting all the things he'd learned in the last few weeks as I spoke. Immediately afterward—with the cameras rolling—I finally got to ask him the question we'd all been dying to know the answer to: Had he changed his mind?

"*Yes*," he said. "How could I not?"

"But," he continued, "how am I going to tell my friends about it?"

He has never met former Kiribati president Anote Tong, but he unconsciously echoed the sentiment of his words: "It is time for the world to wake up and understand: we are all Kiribati."

NO TIME TO WASTE

"Every bit of warming matters. Every year matters and every choice matters."
PETTERI TAALAS AND JOYCE MSUYA, *IPCC SPECIAL REPORT ON GLOBAL WARMING OF 1.5 DEGREES*

"You don't have to know where we'll end up. You just have to know what path we're on."
KIM COBB, A CLIMATE SCIENTIST COLLEAGUE OF KATHARINE'S

The window of time to prevent truly dangerous levels of change is closing fast. Despite what you may have heard, there's no magic number, date, or threshold that will entirely save us from climate change's effects. Trying to put a number on exactly how much global temperature change is "dangerous"—and how much carbon we can put into the atmosphere before we hit that level—is like trying to put a number on exactly how many cigarettes we can smoke before we develop lung cancer.

We know that the more we smoke, the greater the risk. But we also know there's no single threshold before which our health is perfectly fine and after which it's all over. It's not as if we could smoke 9,999 cigarettes and we'd suffer no consequences at all, but if we smoked that 10,000th one, then lung cancer would bloom overnight. With both cigarettes and carbon emissions, all science can say is: the sooner we stop, the better.

As a climate scientist, my research focuses on how climate change affects people, places, and other living things. To do that, we have to calculate how fast, and by how much, climate could change in the future. This isn't a new question. The very first person to ask—and answer—it was Swedish physical chemist Svante Arrhenius. His mother was a Thunberg, making him a distant cousin of a modern Thunberg you may have heard of, Greta.

Arrhenius won the Nobel Prize for chemistry in 1903, but he must have been a curious person, because in the 1890s he was wondering about something that had nothing to do with his work on ionic dissociation. Rather, he was asking, what will happen to Earth's temperature as burning coal increases carbon dioxide levels in the atmosphere? He figured it should be possible to calculate this using the physics and chemistry of the day. So he rolled up his sleeves and got on with it.

In 1895 he presented the results of his research to the Royal Swedish Academy of Science. The next year he summarized them in a study titled, "On the Influence of Carbonic Acid in the Air upon the Temperature of the Ground." By season and by latitude, he calculated how much the planet would warm if humans increased the level of carbon dioxide by 50 percent, 100 percent (a doubling), or 200 percent (a tripling) compared to the 1890s.

Using what scientists understood of the physics of the atmosphere at that time, Arrhenius was also able to calculate how much faster the Arctic would warm than the rest of the planet, exactly as I saw when I visited there with Steve Amstrup and the Polar Bears International team in October 2015. As the polar bears see their world changing around them, so do we; but with one big difference. We're the ones causing it.

At the time Arrhenius was doing his calculations, carbon dioxide levels had only risen by about 5 percent relative to their pre-industrial values of 280 ppm. He thought it would take centuries, maybe even millennia, for them to double or more. But as the Industrial Revolution accelerated, carbon emissions began to increase exponentially. By 1958, when another pioneering chemist, Charles Keeling, began to measure carbon dioxide at the Mauna Loa Observatory in Hawaii, atmospheric levels were at 316 ppm, a 13 percent increase. Today, they're over 420 ppm. That's a 50 percent increase relative to pre-industrial levels. And carbon dioxide is well on its way to a tripling within this century if we humans don't radically alter the trajectory we are on. Today, we know one crucial thing Arrhenius didn't: how urgent this is.

HOW MUCH CARBON IS TOO MUCH?

The more carbon we produce, the faster the climate changes, and the greater the danger for all of us. The reason we can't put exact dates or figures on risks is not because scientists don't have a good idea of what the impacts will be at different levels of carbon dioxide in the atmosphere. It's because different people assess risk differently and are differentially vulnerable to climate impacts.

So, the magic number? It's as low as we can go. As far as we humans are concerned, the perfect temperature for us is the temperature we've already had for the past few thousand years. And the lower our total emissions are, the greater our chances of avoiding the vicious feedback cycles like methane emissions from thawing permafrost. As Petteri Taalas, secretary general of the World Meteorological Organization, and Joyce Msuya, deputy executive director of the United Nations Environment Program, wrote in the foreword to the IPCC's *Special Report on Global Warming of 1.5 Degrees* in October 2018: "Every bit of warming matters. Every year matters and every choice matters."

Even if we meet the most stringent targets of the Paris Agreement, slashing global carbon emissions in half by 2030 and achieving net zero (where the amount of carbon we take up equals what we produce) before 2060, Earth has already warmed by more than 1°C, or nearly 2°F, over the past hundred and fifty years. For some people in some parts of the globe, that's already enough to be "dangerous."

The Native American villagers in Alaska whose homes were judged to be at "imminent risk" almost two decades ago would agree. Some are already being forced to abandon their homes due to the warming they've already experienced. In Europe, if you talked to the family members of some of the seventy thousand people who died in the heat wave of 2003—a heat wave that was twice as likely as a result of a changing climate—they'd probably say the amount of change we've already experienced qualifies as dangerous. In western Canada and southern Australia, many whose homes are being lost to increasingly destructive wildfires would agree—climate change is dangerous now.

But for others, the impacts of climate change on our food, our water, and our safety are only now starting to become concerning. Changes aren't likely to be widespread and persistent enough to pose a serious threat to our lives and livelihoods until we reach the Paris targets of 1.5 or 2°C. For still others, maybe some who live in a higher-latitude location with ample resources but few concerns about permafrost, sea level rise, or wildfire, it might be even longer before the mercury reaches a level that they would consider dangerous—perhaps 2.5 or even 3°C of warming.*

Here's the problem, though. The time to limit the warming isn't when we actually hit whatever target we choose. By then, it's too late. It's like quitting smoking after you've been diagnosed with some scary spots on your lungs. In the same way, we humans have to decide what's dangerous well ahead of time and act now to prevent it. That's why the goal of my own research has been to quantify the damage rising temperatures will inflict on the systems that sustain our lives—water, food, infrastructure, health—so that we can make wise decisions now.

HUMANS ARE THE BIGGEST UNCERTAINTY

The concept that we have a choice to make is surprisingly new. Back in the 1990s, nearly all regional and sectoral climate assessments treated climate impacts as essentially inevitable. They were just looking ahead to see what was going to happen, so people could prepare. By doing so, they cast humans as the archetypal victim in an old western, tied to the tracks while the locomotive steamed around the corner. There's no alter-

---

*Given the interconnected nature of our world and our tendency as scientists to underestimate the extent of climate impacts, I feel this is unlikely. But since I'm a scientist, I have to say it's at least hypothetically possible.

ing the speed of the train, this metaphor suggested, but if you could see it coming at least you could prepare to minimize the impact.

This view isn't just unhelpful—it's wrong. Why? If disaster isn't inevitable, and we can do something about it, understanding the difference our choices make becomes absolutely critical. This simple concept is the key to everything I do, and everything I talk about in this book.

What *are* our choices? As John Holdren, senior science advisor to President Obama, declared in his 2008 address to the U.S. National Academy of Sciences, we have three of them. We can reduce the heat-trapping gas emissions that are causing climate to change; we can build resilience and prepare to adapt to the changes that we can't avoid; or we can suffer. "We're going to do some of each," he said. "The question is what the mix is going to be. The more mitigation we do, the less adaptation will be required, and the less suffering there will be."

To correct the train metaphor, we humans are actually on the locomotive's footplate, with our hand on the throttle. The train is heading for a bridge that's down. We can assume protective positions to ride out the crash, but we can also stop accelerating (stop increasing our emissions) and hit the brakes (decrease our emissions) to minimize the damage. We'll have a lot better chance of surviving, the more we do.

So while localized information on how impacts affect us helps people understand why climate change matters, it's essential to pair this information with an understanding of how our actions matter, how impacts depend on the carbon emissions we produce. This information is time-sensitive: it presents us with a choice to make now. If we don't act, that in and of itself is also a choice; and it's one that makes the worst-case scenario, with all of its attendant suffering, virtually inevitable.

We humans are the greatest uncertainty in the climate system.

THE DIFFERENCE OUR CHOICES MAKE

When the Union of Concerned Scientists invited me to help lead a new regional climate assessment for the state of California in 2002, I knew that we couldn't just use the same mid-range, middle-of-the-road scenarios previous assessments had. We had to show the difference that human choices can make. So we decided to compare two very different climate futures: First, what if humans continue to depend primarily on fossil fuels for the rest of the century and emissions continue to rise? And second, what if we accelerate the transition to clean energy with climate-friendly policies, bending emissions down? What would California look like in those two different futures?

The all-star project team included Steve Schneider, one of the most experienced climate scientists of our time. A physicist who'd served as advisor to U.S. administrations since the Nixon era, Steve's outspoken advocacy over the decades has encouraged many of us scientists to also speak out about the risks of climate change. A California resident, Steve wanted to know the answers to these questions as much as anyone.

The two scenarios were first fed into the complex global climate models that capture the physics and chemistry and, increasingly, the biology of the planet, to see how the Earth's climate system would respond to each possible future. Global models and top colleagues weren't enough, though. Global model output is notoriously coarse and would not give a precise enough picture of the impacts of climate change on California's varied terrain. I had to figure out how to "downscale" the global climate model output into high-resolution, finely gridded information. I called Ed Maurer, a hydrologist who'd recently moved to California's Santa Clara University. "I'm not sure about this," Ed said cautiously, "but it sounds important. Let's see what we can do!"

I collected the temperature and rainfall projections from the global models. Ed gathered the historical observations. Together, we bias-corrected and downscaled them. Then we parceled the data out to the team of over thirty researchers who were studying climate impacts on California's water, wildfire risk, health, wine industry, air quality, and more. For the first time, everyone's analysis was based on the same climate projections, and everyone was quantifying the difference between a high-carbon versus a low-carbon future. And when the results started to trickle in, they were even more shocking than I'd expected.

It turns out that for our modern world, the difference between a higher versus a lower emissions future is nothing less than the survival of our civilization. In the lower emissions scenario, our agriculture, our water, and our economic systems can continue, albeit with significant and often costly adaptations. The higher emissions scenario predicts the end of many of these systems as we know them. For example, the city of Sacramento could experience a Tucson-like climate before the end of the century, and up to 90 percent of California's winter snowpack (which supplies half its water) could disappear.

The results of our work, published in the *Proceedings of the National Academy of Sciences*, hit home. A year later, California governor Arnold Schwarzenegger signed the very first law in the history of the United States requiring mandatory greenhouse gas emission targets. Standing behind him at the ceremony were the California authors of our study. The law, executive order S-3-05, listed the impacts we'd found were projected to occur under the higher scenario as justification for why California was taking action. "I say the debate is over. We know the science. We see the threat. And we know the time for action is now," the governor said.

The message had been received, loud and clear: we have a choice, and now is the time to act.

THE SICKNESS AND THE CURE

"Why do you fight for our planet? For me, it's to save lives, protect our children's future, fight for social and racial justice, and support people's health and human rights around the world."

GAURAB BASU, CAMBRIDGE HEALTH ALLIANCE

"There is no Planet B."

CLIMATE ACTIVIST'S PLACARD

A few years ago, I received an invitation to an event that sounded so improbable I almost deleted the email, suspecting it was a fake. A festival held in the Canary Islands organized by Queen guitarist Brian May and his former PhD advisor, which had featured Brian Eno as a musical guest the previous year . . . could that be real? Fortunately, I decided to look it up before deleting it. I learned that Starmus was indeed a real festival of science communication and art, and they really were inviting me to speak at it. I accepted, and was flabbergasted to end up with a front-row seat at one of Stephen Hawking's final talks.* I'd read the British physicist and Nobel Prize winner's best-selling 1988 book, *A Brief History of Time*, as an astrophysics undergraduate. His grandiose vision of the universe injected meaning and life into the endless and often tedious equations required to understand quantum

---

*Having lived with from motor neuron disease throughout his adult life, Stephen Hawking died in 2018 aged seventy-six.

probabilities and nonlinear fluid dynamics, so seeing him speak had been on my bucket list for years.

Hawking's talk, like his posthumous book, focused on the future of the human race. He repeated his concern that climate change was one of the greatest threats facing this world and stressed the urgent need to avoid catastrophic impacts. "There is no new world, no utopia waiting around the corner," his computerized voice said, its very lack of inflection making the warning even more ominous. I agreed—but my jaw dropped in surprise when Hawking concluded by saying humans will have to populate a new planet to survive climate change. As a climate scientist, I know the speed of climate change far exceeds our ability to terraform Mars. Unchecked, climate change will overwhelm our civilization long before anywhere near the number of people required to "save civilization" could be transported to another planet where they'd thrive. And even if a few hundred or even thousands did manage to make it to Mars, they wouldn't be the poorest and most vulnerable of us, who are already suffering the greatest impacts. It would be the Jeff Bezoses and Elon Musks of the world who could afford to buy their own ships and load them with their favorite people like a real-life version of the black comedy *Kingsman*.

Two days later, I was waiting backstage for my own talk. The speaker immediately before me was Martin Rees—Lord Rees—Astronomer Royal of the United Kingdom, whose groundbreaking work on galaxy clustering and quasars I'd also read as a student. As technicians labeled our identical laptops with different colors of tape to make sure they didn't get confused, I finally asked him the question that had been gnawing at me since Hawking's talk. "Do you agree that we may have to terraform Mars to escape climate change here on Earth?"

"Oh no," he replied categorically. "Stephen and I are old friends. But fixing climate change is a doddle in the park compared to terraforming Mars."

That was a mic drop if I'd ever heard one.

CLIMATE SOLUTIONS ARE HEALTH SOLUTIONS

There is no backup planet in the wings; whether we like it or not, this is our world. That's why, when it comes to our choices, it isn't only about avoiding the worst: it's also about making our planet a better place to live. And nowhere is the contrast between the risk of inaction and the reward of action more evident than when it comes to our health.

Ed Maibach is a runner. He's also a public health researcher who directs George Mason University's Center for Climate Communication. And Ed's research confirms, time and time again, that our health depends on the planet's health. Just as smart individual choices in diet and lifestyle benefit our short- and long-term health, so, too, smart societal choices can reduce both the severity of climate change and its impact on our health.

"Americans tend to see global warming as a distant threat," one of Ed's 2018 articles begins. But he's found that giving people information on how climate change affects their health makes them more engaged and more willing to support climate action. What kind of information? How climate change affects us directly, through heat waves, stronger storms, flooding, and indirectly, through air pollution, disease, and contamination. It affects the quality of our food, the safety of our homes, and even our mental health. And what type of action can we take? First, mitigation: cutting air pollution, heat-trapping gas emissions, and climate change. And second, adaptation: preparing to weather the impacts we can no longer avoid.

"The health benefits of climate solutions," he told me, "are profound, nearly immediate, and local—which in turn helps to address the psychological conundrum of climate change being perceived as distant. Climate solutions *are* health solutions; and not just in the distant future, but today, here, for us. They pay for themselves almost immediately, so their future benefits are essentially free."

BEATING URBAN HEAT AND AIR POLLUTION

When we think of climate change and health, our minds often jump right to the most obvious connection: stronger and more frequent heat waves. And these are a big deal. As I mentioned before, in July and August 2003, the then hottest summer on record in Europe caused more than 70,000 premature deaths across the continent. Even back then, climate change had already doubled the risk of such a heat wave occurring. And as the mercury ticks up, tempers also flare. Psychologist Craig Anderson has been studying this phenomenon since the 1980s. "Hot temperatures increase aggression by directly increasing feelings of hostility and indirectly increasing aggressive thoughts," he wrote in a 2001 paper called "Heat and Violence." That's why warmer cities have higher rates of violent crime than cooler ones, and violent crime tends to spike in the summertime.

A national report from the Union of Concerned Scientists found that if emissions continue unchecked, by midcentury the U.S. will see twice as many days with a heat index above 38°C or 100°F as today, and four times as many days with a heat index above 41°C or 105°F. "Failing to reduce heat-trapping emissions would lead to a staggering expansion of dangerous heat," the authors wrote. This heat would be worse in cities, thanks to the urban heat island effect—created by high concentrations of concrete, asphalt, and buildings in cities. It's also worse in poorer neighborhoods, which are less likely to have many trees. Trees cool their local environment by releasing water vapor and providing shade. And the more air-conditioning we use, powered by fossil fuels, the higher the emissions and the hotter the planet will get. It's a cruel feedback loop.

Then there's the fact that burning fossil fuels doesn't only produce heat-trapping gases, it also generates air pollution. Air pollution from burning fossil fuels is responsible for nearly 9 million premature deaths *per year*. To put those horrifying numbers into context, by spring 2021 there had been 3 million deaths from COVID-19, worldwide. That's why the World Health Organization (WHO) has called air pollution the "single largest environmental health risk" facing humanity; and climate change only makes it worse. Higher temperatures speed up the formation of hazardous ground-level ozone that forms from tailpipe and industrial emissions. Ozone, particulates, and other pollution makes it hard to breathe deeply, causes coughing, and damages our lungs, leaving them vulnerable to infection. Those who suffer from lung conditions such as asthma, emphysema, and bronchitis are especially at risk, but high ozone days can harm even healthy people. And air pollution exacerbates coronavirus, by making people's lungs more vulnerable to the infection.

In the Netherlands, researchers found that even a 20 percent increase in exposure to particulate pollution doubled the risk of COVID-19 infection. A U.S. study found that people living in polluted areas were much more likely to die of the virus. Analyses in China, Italy, Europe, and elsewhere found similar results. Not only that, but those who were already impoverished or disadvantaged were more likely to have been exposed to more air pollution prior to the pandemic, worsening their vulnerability. In Chicago, Illinois, African Americans make up less than a third of the population but more than two-thirds of COVID-19 deaths; Harvard researchers believe that air pollution may account for some of that disparity. And just as scary are new findings on how air pollution harms our brains. It can increase the risk of dementia and other neurodegenerative disorders that so many older adults experience. It can also affect the newly developing brains and nervous systems of babies before they are born, delaying or impairing cognitive development and increasing risk for autism.

At this point you might be tempted to assume the fetal position yourself. How are we supposed to prevent such hideous, pervasive, yet invisible harms? Ed has the answer to that: implement solutions that help today, and tomorrow. For example, tackling the urban heat island effect, the very factor that makes extreme heat and air pollution worse in urban centers, can reduce climate impacts and make us better prepared to weather them.

Take the city of Chicago, for example. It's classified as a "severe ozone nonattainment area" by the Environmental Protection Agency due to its endemic air pollution. It also experienced deadly heat waves, responsible for hundreds of deaths, in 1995 and again in 1999. In 2008, the city's mayor announced an ambitious Climate Action Plan for Chicago. In it, the city identified the most vulnerable people and neighborhoods and worked with me and other climate scientists to determine how these risks were likely to change under a higher and lower emissions future. For example, the city's first responders told me that they staffed by the thermometer in the summer. As heat spiked, health emergencies, violence, and crime in the South Side of the city rose a noticeable amount. Then they identified key things they could do today that would make people's lives better now, reduce carbon emissions, and ensure the city did its part toward achieving a lower emissions future. To mitigate summer heat, they identified "hot spots" where trees, "green roofs" planted with vegetation, and reflective surfaces would cool temperatures during the hottest days. This would slow down the reactions that create ozone pollution and reduce energy use and carbon emissions at the same time.

How are they doing today? As of 2019, Chicago has hundreds of green roofs around the city, including an expansive one on its city hall where 150 species of plants now grow. They've shut down the two local coal-fired power plants that were a big source of air pollution and carbon emissions. They are well on the way to building one of the largest electric bus fleets in the U.S. and have increased the number of bicycle trips people take to over 45 million each year. There's still a ways to go in meeting their goal of six thousand green roofs and a million trees, but they are heading in the right direction.

Heat isn't their only problem. More frequent and more intense heavy rain events have been flooding the city, shutting down the public transportation system. It's also leading to sanitary sewer overflows into the Chicago River and beach closures on Lake Michigan. Flooding is so severe that, in 2014, Farmers Insurance sued the county and the city's water district, stating that they "knew or should have known that climate change . . . has resulted in greater rainfall volume, greater rainfall intensity and greater rainfall duration." As a result, the city has built two new reservoirs to hold the storm water. And most importantly, everything they've done has helped cut costs, clean up the air people breathe, make the city a safer place to live, and reduce carbon emissions. It's a win-win-win.

COMBATTING DISEASE

Warmer air holds more water vapor, allowing storms to pick up more moisture. This increases the risk of heavy rain events and exacerbates floods. Floods pick up even more unpleasant stuff and it all gets washed along—into our streets, our homes, our water supply, and our drinking water. It's not only happening in Chicago. It's even worse in the developing world, where floods can lead to outbreaks of deadly waterborne illnesses like cholera and typhoid fever, thanks to sewage overflow and contaminated drinking water supplies. Every year, some 3 million people across the globe already die due to waterborne illness, which can be caused by bacteria, viruses, or parasites. Diarrhea alone kills 2 million people, a quarter of whom are children under five, according to the WHO.

Then there's the fact that hurricanes, typhoons, and cyclones are getting bigger and stronger, with a lot more rainfall. In 2017, Hurricane Harvey powered up from a tropical depression to a category 4 storm in a mere forty-eight hours, thanks to record high temperatures in the Gulf of Mexico. It was likely responsible for nearly 40 percent more rainfall and four times the economic damage compared to what would have happened without climate change. The storm is estimated to have caused sixty-eight deaths directly—but it also flooded dozens of wastewater treatment plants across southeast Texas, releasing over 30 million gallons of sewage into streets and waterways. The sludge increased everything from skin infections to diarrheal disease and left many homes with nightmarish mold problems.

The year before, in 2016, Hurricane Matthew caused forty-three deaths and billions of dollars of damages across many Southeast U.S. communities, including Princeville, N.C. But when the same hurricane hit Haiti, a poor country that already suffered from lack of infrastructure, it was orders of magnitude more devastating. Hurricane-related flooding caused a significant jump in cholera cases as floodwater contaminated wells and drinking water. Crops in some areas were nearly completely destroyed, along with over two hundred thousand homes, creating a humanitarian crisis. WHO flew in a million doses of anti-cholera vaccines, but for many it was already too late.

We often take for granted that we have a tap to turn on, let alone that when we do, clean drinkable water will come out. Around the world, 2.2 billion people (that's almost 30 percent of the world) lack safe drinking water, and 4.5 billion—over half the world—don't have access to safe sanitation that prevents their excrement from contaminating their local environment. Ensuring clean water and sanitation for all is one of the U.N.'s Sustainable Development Goals. "Water can help fight climate change," the goals remind us, and "everyone has a role to play. There are sustainable, affordable and scalable solutions" available today. Like what?

Sanergy is an organization that builds toilets in Kenya's urban slums and schools. They carefully and safely collect the waste, so it doesn't get into people's food or water and make them sick. Then they turn the waste into fertilizer and animal feed that increases crop yields and can replace commercial fertilizers that produce nitrous oxide, a powerful heat-trapping gas. Others are even turning waste into fuel: Sulabh International, the largest nonprofit organization in India, has built nearly two hundred plants that use the waste from public toilets to create renewable natural gas, or biogas. It can replace fossil fuels in cooking and electricity-generation. And in Grand Junction, Colorado, they're doing the same: turning human waste into biogas that fuels the city's garbage trucks, street sweepers, and buses.

Biogas isn't the only way hurricanes and flooding can spur clean energy investments. In 2017, Hurricane Maria ravaged Puerto Rico. It's estimated to have caused several thousand deaths, while also destroying more than 80 percent of the territory's utility poles and transmission lines. Storm damage caused the longest blackout in U.S. history—in some places, over eleven months without power. For many hospitals and senior citizen residences, this was a key contributor to the mounting death toll. In 2020 the Puerto Rico Energy Bureau issued its first order to start building the solar and battery capacity that will ultimately transition the island to 100 percent clean energy. This slashes its carbon emissions, cuts its electricity costs,* and makes its energy grid more resilient and better prepared to weather the impacts of stronger, climate change–fueled hurricanes. Again, it's a win-win-win.

PREVENTING REFUGEE CRISES

Then there's the fact that climate change has the potential to exacerbate existing conflicts and refugee crises, and even spark new ones. These can have devastating impacts on human health and well-being. Without access to food, water, and health care, things we take for granted like feeding our family or dealing with an infection can rapidly become life-threatening situations.

The United Nation's Global Compact on Refugees, adopted in 2018, states plainly that "climate, environmental degradation and natural disasters increasingly interact with the drivers of refugee movements." It's not that climate change is causing most of these events: rather it is taking already precarious situations fraught with resource scarcity, civil or political conflict and instability, poverty, and hunger, and making them worse. An Oxfam report estimates that climate-fueled disasters are already displacing some 20 million people each year, the majority of whom live in Asia. For countries already nearing the brink of disaster thanks to social, economic, political, and cultural stresses, climate change impacts could be the final straw.

While many countries have opened their borders to refugees, a better solution would be for them not to have to flee in the first place. Programs designed to alleviate poverty can increase people's resilience to climate impacts. Subsistence agriculture often involves clearing and burning tropical forests. These contain massive amounts of carbon that are released into the atmosphere when they're burned, between twelve and thirty-five thousand tons per square kilometer or from thirty to

---

*Before Hurricane Maria, Puerto Ricans were already paying twice as much for their electricity, on average, than the average American living in the lower forty-eight states.

ninety thousand tons per square mile. So it was very good news when researchers found that a poverty alleviation program in Indonesia unintentionally reduced deforestation by 30 percent at the same time. Other programs in Nepal, China, and India allow communities to manage their forests, which also reduces poverty and deforestation while protecting the resources the community depends on for food and fuel. And intact forests provide safe habitat for animals, which in turn decreases the risk of zoonosis, the process by which viruses like SARS-CoV-2 jump from animal to human populations. Yet another win-win.

MAINTAINING OUR MENTAL HEALTH

Finally, climate change doesn't only affect us physically; it affects our mental health, too. The Oxford Dictionary defines *eco-anxiety* as "extreme worry about current and future harm to the environment caused by human activity and climate change." It adds that it is not considered to be a mental disorder since it is a "rational response to current climate science reporting." I understand; I don't read "cli-fi" (fiction based on apocalyptic climate scenarios) for the same reason. Reality is bad enough; I don't need more of it.

Use of the word "eco-anxiety" jumped a stunning 4,000 percent in 2019 as many young people reported increasing feelings of anxiety, panic, and fear related to climate change. "If you've heard about the grim climate research," psychologist Britt Wray says, "you've probably felt bouts of fear, fatalism or hopelessness. If you've been impacted by climate disaster, these feelings can set in much deeper, leading to shock, trauma, strained relationships, substance abuse and the loss of personal identity and control."

Yet when I'm talking to students and young people, and hearing from psychologists who study despair, a common theme emerges: acknowledging these emotions is the first step to action. People often tell psychologist Renée Lertzman, "I care very deeply about what's happening, but I feel like my actions are insignificant. And I don't know where to start." But accepting who you are and what you feel *is* a first step.

The next is to connect with others, support one another, and raise your voices together. Many young people say it's their anxiety that led them to look for, participate in, and even lead advocacy campaigns, from Greta Thunberg's Fridays for Future climate strikes to Jamie Margolin's Zero Hour. Again, impacts and solutions go hand in hand.

CLIMATE CHANGE IS OUR BOTTOM LINE

Climate change touches every single one of the issues that fill the head-lines: public health concerns, food security, humanitarian crises, resource scarcity, the economy, and the impact of disasters on our cities and infrastructure. I've only described the very basics of how climate change affects our health, and how many inspiring, positive solutions there are that can better people's lives today and ameliorate climate change tomorrow. But there's one more important point to make.

What do all these impacts have in common? While they affect every single one of us, it's the most vulnerable, most marginalized, and most disempowered people who tend to be hurt first and worst. This is true whether it's heat waves in New Delhi or the South Side of Chicago, air pollution in Shanghai or Los Angeles, floods in London, or Bangladesh. Those most at risk are those who have already lost the most. And it's no accident that these are the same people who were most harmed by the coronavirus pandemic and by its economic impacts. The 2020 *Lancet* Countdown, which tracked the connections between climate change and public health, concludes, "We don't have the luxury of tackling one crisis at a time. COVID-19, climate change, and systematic racism represent converging crises that need to be tackled in unison."

Here's the good news, though. Through his research, Ed Maibach and his colleagues have found that people's awareness of the health risks of climate change has increased significantly. Back in 2014, only around 30 percent of Americans thought the health impacts of climate change would increase. By 2020, well over 50 percent of people surveyed in the U.S. agreed that the impacts of extreme heat, harm from stronger storms and hurricanes and more dangerous wildfires, and impacts on our lungs from air pollution, allergies, wildfire smoke, and pollen were on the increase.

The bottom line is this: climate change is not only a science issue. It is not "just" an environmental issue. It is a health issue, a food issue, a water issue, and an economic issue. It's an issue of hunger, and of poverty, and of justice. It's a *human* issue.

By following this train of thought, we arrive at a simple yet potentially revolutionary realization: getting people to care about a changing climate doesn't require them to adopt "new" values. Gone is the burden of inspiring people to "care" about deforestation and melting ice caps. No need to teach them to hug a tree, respect a polar bear (hugging not advisable), or throw themselves into recycling. And good-bye to partisan divides. As Ronald Reagan stated in 1984, "Preservation of our environment is not a partisan challenge; it's common sense. Our physical health, our social happiness, and our economic well-being will be sustained only by all of us working in partnership as thoughtful, effective stewards of our natural resources."

We humans are the reason why climate is changing, but that also means our future is in our hands. This is why Steve Amstrup and his team of polar bear scientists are so focused on telling people about the threats posed by global warming and what we can do about it. This is why Steve Schneider, who died in 2010, fought so fiercely for the planet we all call home.* And this is why I'm so focused on communicating the risks of a changing climate myself.

Together, we confront both a challenge and a hope. Although some impacts are already here today, the future is yet to come. It's still possible to save the polar bears—and ourselves. As Christiana Figueres says, "The full story has not been written. We still hold the pen."

So let's talk about how we might do that next.

---

*His memoir is called, appropriately, *Science as a Contact Sport: Inside the Battle to Save Earth's Climate*.

SECTION 4:

WE CAN FIX IT

WHY WE FEAR SOLUTIONS

"Denialism . . . keeps at bay what might be fears, guilt, and a sense of shame, not to mention all that lurks behind a need for CO_2-belching markers of identity such as wait out in the car park."

ALASTAIR McINTOSH, *RIDERS ON THE STORM*

"I know the EPA is just making all of this up to take away my wood-burning stove."

COLLEGE INSTRUCTOR TO KATHARINE

Fear of government encroachment in people's lives is rampant across much of the U.S. and particularly in Texas. Where I live, when they have a primary election, there's a Republican candidate and then a *conservative* Republican candidate who campaigns on the premise that the first candidate is just a government shill. During the coronavirus lockdown, a Texas hairdresser made national news in the U.S. by reopening her salon in defiance of state and county orders. When the county judge sent her to jail, the governor amended his own lockdown order to get her released. A statewide mask order wasn't put in place until long after the state repeatedly struck down local ordinances. This fear of any kind of regulation is a major obstacle to taking action on climate change as well.

When I was speaking to a group of water managers in southeast Texas a few years ago, I put everything I'd learned about avoiding science overload, decreasing psychological distance, and communicating urgency into practice. I didn't start by rattling off ice core data: I began by talking about our shared value—water, in this case, how here in Texas we either have way too much of it or not enough.*

They'd had to cope with a number of challenges in recent years, including a severe drought and multiple floods. I carefully connected water and climate, then climate and carbon, using simple analogies that made the information clear and relevant. But I couldn't avoid the conclusion that the more carbon we pump into our atmosphere, the greater the risk of serious and even potentially dangerous consequences—consequences that will become increasingly impossible to prepare for and adapt to.

At the end of my talk, an older man raised his hand. He cleared his throat. And then, in a reasonable and genuinely concerned tone, he stated:

"Everything you've said makes sense. But I don't want the government telling me how to set my thermostat."**

He seemed accepting of the science, and even concerned about the impacts—but allergic to what he perceived to be the solutions.

WE DON'T THINK WE MATTER

This man is not alone in feeling this way. My home country of Canada has the reputation of taking its climate commitments seriously. It supports the lower 1.5°C target of the Paris Agreement, it has a price on carbon, and in the last election every major political party had a climate plan.*** I've spoken with ministers whose portfolios range from se-

---

*The running joke in many parts of the state is that we get twenty inches of rain a year—all in one day.

**He may have been referring to Jimmy Carter, who, during the seventies oil crisis, famously advised people "to put on a sweater"—and was subsequently pilloried by conservatives who didn't want the President fiddling with their thermostat.

***These plans were of varying quality. I scored them on ambition and feasibility with my colleague, economist Andrew Leach from the University of Alberta. They ranged from an A+ for the Green Party's ambition to an F for the feasibility of the Conservative Party's plan.

nior citizens to infrastructure. They're all concerned about how climate change is affecting their responsibilities and their constituents. Yet the number one objection I regularly hear from fellow Canadians is "We're responsible for just 2 percent of global emissions. Why should we have to fix this? It won't make a difference anyways and it will just harm our natural resource–based economy."* The Liberal government that put a price on carbon partially agrees: they've funded a controversial oil pipeline to take the oil from Alberta's tar sands to ports on the west coast, with the goal of shoring up that province's economy and placating its industry while providing funds to accelerate a green transition and a just transition for energy workers.

Back in 2000, Oregon sociologist Kari Norgaard decided to study why and how people reject climate action. She didn't head down to Texas or up to Alberta, though. Instead, she moved to Norway, another northern country with a similarly climate-friendly reputation.

Norway leads the world in electric vehicle usage; more than 10 percent of cars on the road there today are electric or plug-in hybrids and this number is growing every year. More than half the cars sold there in 2020 were zero-emission vehicles. Thanks to its oil and gas resources, its sovereign wealth fund is the largest in the world, currently estimated at around $1.2 trillion. In 2019, the Norwegian parliament voted to divest $13 billion worth of their investments in coal, oil, and gas companies that weren't taking action on climate change.

The winter Kari spent in the community she gave the alias "Bygdaby" was unusually warm. People were unable to cross-country ski or ice fish like they normally would. The Norwegian media very clearly linked the warm winter to human-caused climate change. During her time there, Kari interviewed dozens of Norwegians. As she chronicles in her book *Living in Denial: Climate Change, Emotions, and Everyday Life*, she found that people knew a lot about the issue: lack of information or facts wasn't

---

*Never mind that, along with Americans, Australians, and Saudi Arabians, Canadians have some of the highest per capita carbon emissions in the world, and the country is number nine on the list of top ten cumulative carbon emitters since 1750.

the problem. But to her surprise, rather than engaging with solutions, they avoided thinking or talking about climate change. And when asked why, they'd reply, similar to what I've heard from so many of my fellow Canadians, "Norway is such a little country. Why would anything we do make a difference?" They perceived solutions to be onerous, personally harmful, and accomplishing little. So even during an unusually warm winter, their fear of solutions outweighed their fear of the impacts.

WE THINK THE CURE IS WORSE THAN THE DISEASE

Our collective threat meter is unbalanced. In fact, sometimes it's tipped all the way in the wrong direction. Even people who agree that climate is changing due to human causes still see the impacts as distant and far off. But that's only half the problem. The other half is that they view the threat from potential climate *solutions* as imminent. They believe government's and society's attempts to address climate change will decrease their quality of life, pummel the economy, and compromise their personal rights.

The concept that much of the resistance to climate change is really a rejection of what people perceive to be unpleasant or unpalatable solutions is known as *solution aversion*. This counterintuitive term was first applied to climate change by social science researchers Troy Campbell and Aaron Kay. They noted, as had others, that the big differences in opinion on climate change between Republicans and Democrats in the U.S. were motivated by their political affiliations. Through experimenting with people's reactions, they found that "the source of this motivation [i.e., the negative attitudes of more conservative voters] is not necessarily an aversion to the problem, per se, but an aversion to the solutions associated with the problem." In other words, Republicans didn't have a problem with climate *science* (though they might think they did); they had a problem with climate *solutions*.

After one of my talks at a local college, an instructor followed me out to the parking lot. I had another meeting to get to, so I couldn't stop. But he didn't either. As I pulled out my keys and put my bag in the back, he continued to voice his scientific-sounding objections: I hadn't addressed how the Sun affects climate. Everyone knew that was the dominant cause of climate change, so why wasn't I being honest about it?

In my talk, I had been clear that the Sun's energy has been decreasing over the last forty years, so it can't be the Sun causing us to warm. "And you can't say it's cosmic rays, either," I said, "because they're heading in the opposite direction from climate as well."

"That's not true," he retorted, and, without even taking a breath, continued, "and I know the EPA is just making all of this up to take away my wood-burning stove."

What do changes in solar energy and cosmic rays have to do with wood-burning stoves? Factually, very little; but to his brain, *everything*. Metaphorically speaking, the circuits in our brain that register fear of what we might lose as a result of climate solutions build a direct connection to the circuits that say it isn't real. Why? Because saying "it's not real" is our *defense mechanism*. Admitting that climate change is real and harmful but you don't want to do anything to address it makes you the "bad guy," and who wants that? As I've said before, most of us want to believe we're a good person.

So instead, someone might say, "the climate has always changed," or "the Sun's at fault, and humans have nothing to do with it," or, my personal favorite, "those climate scientists only say this because it's making them rich." With that last one, they're not just rejecting the label of bad guy themselves, they're pinning it firmly on the scientists. We're the *real* villains, alarming everyone for no reason other than our personal gain!

All too often, if we view climate solutions as harmful, we throw up scientific-sounding smoke screens to obscure our real objections. In reality, they have nothing to do with science, and everything to do with ideology and identity. It's our (often subconscious) defense mechanisms compelling us to prove to ourselves that what we believe is true. That in turn makes us the good person and those venal scientists, corrupt leftist politicians, and greedy green energy tycoons the baddies.

This defense mechanism stems from our very basic human need to feel justified in what we believe and who we are. And it also explains why people are so willing to go out of their way to attack climate scientists they've never met before. Again, this is yet another harmful zero-sum game: if you can convince yourself it's someone else's fault, you imagine this will somehow make you feel okay about it at their expense. It may, short-term—but the effect doesn't last long. That's why many Dismissives are so combative; they're constantly in search of confirmation. Arguing, perversely, supplies it.

FINGERPRINTING THE BIGGEST CULPRITS

When it comes to sharing the responsibility for climate change, it's true that we are all part of the problem to some degree. Carbon emissions didn't just magically appear out of nowhere—and those of us in rich countries produce far more than our fair share. At the same time, however, climate change can't be solved by individual action alone. We need collective action, and that means changing the system. Getting hung up on our personal guilt—or resorting to denial to deal with it—is unproductive. Both approaches discourage those who have little power to effect change and ignore those who do.

Some of this solution aversion is naturally occurring, like the water manager who said so perceptively that he didn't want the government setting his thermostat. But there is evidence much of it is deliberately manufactured by people and industries who, though in a small minority in terms of the number of people they represent, are disproportionately influential due to their wealth and power. They have good reason to fear solutions that may significantly impact their bottom line and even their long-term viability.

Who are those who bear greater responsibility and therefore have the ability to effect great change? According to the *Carbon Majors Report* produced by the Colorado-based Climate Accountability Institute, one hundred fossil fuel companies have been responsible for emitting 70 percent of the world's heat-trapping gases since 1988. And even more tellingly, the top eight of them—in order: Saudi Aramco, Chevron, ExxonMobil, BP, Gazprom, Royal Dutch Shell, National Iranian Oil

Co., and Petroleos Mexicanos—have accounted for almost 20 percent of global carbon emissions from fossil fuels and cement production since the Industrial Revolution. Not only that, but most of the eight top the list of the world's richest corporations as well. They've gotten rich at the expense of everyone who's being impacted by climate change—and at least some of them want to keep it that way.

The homepage of the grassroots project Exxon Knew is crystal clear: "Exxon knew about climate change half a century ago. They deceived the public, misled their shareholders, and robbed humanity of a generation's worth of time to reverse climate change." Each of those statements is based on an exhaustive compilation of Exxon's internal memos, emails, reports, and publications. Some of them were even written by me: my master's thesis looked at how reducing methane and other non-CO_2 heat-trapping gases could contribute to international targets. Some of that research was supported and coauthored by Exxon scientists. They all knew what was what; so I wasn't surprised when one 1979 Exxon Petroleum Department report the project cites reads like a summary of this book. "The CO_2 concentration in the atmosphere has increased since the beginning of world industrialization," it says. "The increase is due to fossil fuel combustion [and] the present trend . . . will cause dramatic environmental effects before the year 2050."

Science historian Naomi Oreskes has dedicated her life to researching exactly what large companies like those selling tobacco and fossil fuels knew about the risks their products posed versus what they said in public about it. In her eye-opening 2010 book, *Merchants of Doubt: How a Handful of Scientists Obscured the Truth on Issues from Tobacco Smoke to Global Warming*, Oreskes and coauthor Erik Conway name the scientists and spin doctors who denied the connection between smoking and lung cancer, and show how companies like Exxon and Chevron brought them on board when the threat of climate action first appeared on their horizon.

Using tried-and-tested strategies that Canadian public relations expert Jim Hoggan skillfully dissects in his 2009 book, *Climate Cover-up: The Crusade to Deny Global Warming*, these organizations and their hired guns deliberately sowed doubt into the public discourse about science we've understood since the 1800s. Full-page ads in prominent newspapers, fake "grassroots" campaigns, dark money–funded think tanks to promote bought-and-paid-for experts, legal firms to attack climate scientists to scare and silence them, donations to politicians at every level across the political spectrum: even still, what these companies have spent to stop climate action is only a tiny fraction of what they have to lose if no one's buying fossil fuels anymore.

THE POWER OF MISINFORMATION

Here's just one example of how effective this disinformation campaign can be. In the U.S., the Yale Program on Climate Communication has been tracking people's opinions of the Green New Deal. This is a congressional resolution introduced by Democratic representative Alexandria Ocasio-Cortez of New York and Democratic senator Edward Markey of Massachusetts. It lays out how to transition the U.S. off fossil fuels while ensuring a just transition for disadvantaged communities and those whose basic livelihood depends on the industry.

That doesn't sound like a bad thing, and when it was first proposed in December 2018, there was broad support for it across the political spectrum. This included 57 percent of conservative Republicans and 75 percent of moderate Republicans. But then the manufactured denial swung into play. Politicians and pundits began to paint it as a socialist plot to destroy America. By April 2019, just four months later, conservative Republican support had dropped to 32 percent, moderate Republican support to 64 percent, and even moderate Democrats' support had dropped a few percentage points. Had the Green New Deal changed? No. So what had? Peoples' exposure to a barrage of negative messaging about it, that's what.

Sometimes, people set up a straw man argument instead. You pretend that someone made an outlandish claim so that you can attack both the claim and the person. This is frequently employed by those who have the most to gain from stonewalling climate action. But when someone sent me a photo of a spray-painted billboard claiming that U.S. senator and known climate advocate Bernie Sanders wanted to abort all the babies to fix climate change, they'd gone too far, I thought. That's so ridiculous, surely no one would believe that. But, as so often occurs these days, they could and they did.

A few months after I'd seen that photo, I received an invitation to speak at a religious university so conservative—and not just theologically, but politically and culturally, too—that I was shocked they'd even let me in the door. The invitation came from a fellow scientist and colleague named Evan, whom I'd met at various conferences over the years.

A week before the talk, Evan emailed me.

"I just wanted to let you know what some are saying about your visit," he said, attaching several emails he'd received from other faculty and administrators.

As I scrolled down, my eye snagged on a quote from one professor: "This is the work of Satan, the father of lies . . . presenting solutions to climate change is morally equivalent to abortion." An administrator joined in, telling Evan he had to stop advertising my talk. It was making too many people angry.

Yup, they'd bought it.

TURNING OPINIONS AROUND

I needed to reconsider what to say. The stakes were high. I didn't just have to talk about why climate change matters to us as Christians. I also had to show there were solutions that helped rather than harmed people and that had nothing to do with abortion and were compatible with our values. I wanted to do the opposite of what had been done with the Green New Deal: start from a negative opinion and turn it right around. But how?

I found myself thinking of Mitch Hescox. He's a coal executive turned pastor who now leads the Evangelical Environmental Network. He's passionate about helping people who live in areas similar to his home in Pennsylvania, where the coal industry shores up the local economy while simultaneously poisoning the community's water and its air. Mitch has seen firsthand the devastating impacts fossil fuels have on our health. He understands that the consequences are particularly bad for children, pregnant women, and unborn babies. As I talked about in the last chapter, fossil fuel use and its cascading impacts on heat, pollution, water, food, disasters, and security affects our health in a myriad of ways. That's why Mitch refers to climate change as a "pro-life" issue. If Christians are truly pro-life from conception to death (rather than from conception to birth, as some people's attitudes seem to suggest), they should be leading the charge to get rid of fossil fuels—not dragging their feet at the back or heading in the other direction.

At the same time, though, it's also necessary and appropriate to acknowledge the benefits that energy has brought us. Growing up as a missionary kid in Colombia was an experience I knew many at that university could relate to. Anyone who's spent time in a poor country knows how profoundly poverty and lack of access to resources can impact people's lives; but they might not know how directly fossil fuel extraction exacerbates it. In Colombia, for example, coal mining in the impoverished and largely indigenous department of La Guajira is, as political scientist Noel Healy puts it, "connected to a system of production entrenched in violence, bloodshed and environmental destruction." Just one example is the coal industry's exorbitant water use during a recent climate-amplified drought. This led to massive shortages that left whole villages without water.

From rural Appalachia to the oil- and gas-rich Niger Delta, pregnant mothers, babies, and children face increased risk of cancer and birth defects due to fossil fuel extraction. Emem Edoho, an advocate for Nigerian children, says, "Oftentimes, residents . . . live in agonizing conditions, economically and socially, arising from years of neglect and deprivations and severe environmental degradations, caused by the production activities of most oil and gas companies in the area." These environmental harms aren't offset by financial gain, either; income inequality in the developing world is actually exacerbated by fossil fuel extraction, as kleptocrats find a way time and again to line their own pockets with the profits.

Then there are the health problems associated with the combustion of fossil fuels. As I mentioned earlier, air pollution from fossil fuels is responsible for more than nearly 9 million deaths worldwide each year. (There are also several million deaths each year due to exposure to open indoor fires that many impoverished women have to cook on.) Two hundred thousand of these annual air pollution deaths are in the U.S. Who are these people suffering from lung disease, asthma, and other related conditions? Often those who can't afford to live in a cleaner neighborhood with better air quality, as well as children and those who are already sick or infirm.

Only after I'd thoroughly explored all of these issues did I turn to climate change and how the suffering global warming will cause around the world is not going to be parceled out equally. One Stanford study estimates climate change has already increased the economic gap between the world's richest and poorest countries by as much as 25 percent. It has also negated over fifty years of advances in poverty and hunger reduction and could push 120 million more into poverty by 2030. The global south will continue to bear the brunt of future impacts, with the poorest 40 percent of countries experiencing a 75 percent drop in average income by end of century.

Taken individually, any one of these reasons should be enough to convince us that there are better options than fossil fuels for the developing world, which is already suffering the impacts of climate change. Taken together, they overwhelmingly demonstrate how, if we care about life at all, then we already care about climate change.

So what are the solutions? One of the most surprising is educating and empowering women and girls. Education reduces infant mortality, increases equality, and allows women the freedom to choose how many children they have.* In Kenya, for example, women farmers who were taught about agricultural technology and agribusiness increased their income by an average of 35 percent. In Mali, women who attend second-

---

*In contrast to Catholics, conservative Protestants, like those in my audience at the university, are not generally opposed to birth control.

ary school go on to have an average of three children; women who aren't educated, seven. For each additional year of schooling a mother has, the chance of her child dying before the age of five decreases by 7 to 9 percent on average across developing countries; this increases to 10 percent in Malawi and 17 percent in Uganda. And around the world, children born to mothers who can read are 50 percent more likely to survive childhood than those whose mothers can't. Education, empowerment, and poverty eradication, not abortion, are the positive routes to addressing climate change and building prosperity in the developing world.

As Christians, I concluded my talk, our response to any challenge should be characterized by love. Jesus says, "By this everyone will know that you are my disciples," and the apostle Paul amplifies this, instructing his readers that "the only thing that counts is faith expressing itself through love." Love is key to acting on climate: caring for the poor and the needy, those most affected by the impacts of a changing climate, as well as creation itself. It's not only our responsibility, *it's who Christians believe God made us to be.*

As I finished my talk, I was overwhelmed with enthusiastic students approaching me. Many of them asked, "Here's what I'm studying; how can I apply what I'm learning to help with this issue?" Others shared how they'd never made the connection before, but it was very clear to them now why they cared.

As I walked out, the administrator—he of the email—grabbed my hand and shook it vigorously. He thanked me for the talk. And the next week, Evan forwarded me an email he'd received from him. It said that I'd given him a new perspective on this topic and he really appreciated that Evan had invited me. He understood how his existing values connected with addressing climate change and he'd changed his mind.

This was one of the most encouraging responses I'd ever heard. If we are going to fix this thing, we need everyone to get involved.

CARBON AND
THE COMMON GOOD

"We . . . no longer feel in control of our everyday lives. As we retreat to smaller circles of kith and kin, the Commons goes to seed."

ERIC LIU, *YOU'RE MORE POWERFUL THAN YOU THINK*

"Everyone was surprised how many planets we'd need to support our lifestyle."

MIDDLE SCHOOL GYM TEACHER, AFTER ATTENDING ONE OF KATHARINE'S TALKS

Back when the number of humans could be measured in hundreds of thousands, even millions, the planet was, for all intents and purposes, infinite. Its atmosphere would absorb any pollution, its water was plentiful, its oceans were full of fish, and if food stocks started to run out or land area became insufficient, there was usually more to be discovered.

Over time, the planet stayed the same size. Humans, on the other hand, expanded exponentially. Our consumption habits, particularly in rich countries, expanded even more. Today, nearly every acre of arable land is already parceled out, or home to increasingly endangered ecosystems and species. Water supplies are already insufficient in many parts of the world and are becoming increasingly tenuous as our groundwater withdrawals increase. Overfishing has already led to the collapse and near disappearance of many key food stocks, from Newfoundland's cod to Atlantic bluefin tuna. Occasionally we can roll back the clock: the demand for whale oil for lamps drove whales nearly to extinction before they were saved, ironically, by fossil fuels and electricity. But increasingly, those stories are few and far between compared to those that illustrate, for the hundredth time, our mismanagement of our commons.

THE TRAGEDY OF THE CLIMATE COMMONS

It might be hard to picture something as big as our planet as a shared commons, but this simple metaphor explains many of the challenges of our global resource problem today. It comes from the idea that, during the Middle Ages, many English villages had a common area, or green, where local small farmers grazed their livestock. The best way for an individual farmer to use the green to maximum benefit would be to graze as many animals as possible on it. But if everyone did that, the land would be overgrazed and soon would yield no grass for anyone. As these English farmers recognized, shared land must be grazed according to the common good, limiting the number of beasts each farmer grazes by mutual consent to ensure the land continues to provide for all.

The fundamental concept of a "commons" as a shared resource was introduced by economist William Forster Lloyd in 1833. It wasn't popularized, though, until 1968 when another economist, Garrett Hardin, coined the phrase "*tragedy of the commons*" to describe what would happen if people exploited a shared resource such as common grazing land guided only by self-interest. His point was that our planet is a similarly shared space, but that we had failed to recognize that it also has limits.

Both men used this useful concept to draw some extremely offensive conclusions regarding optimal strategies to manage the commons, however.* It took fifty more years and a woman to prove them wrong. In 2009, economist Elinor Ostrom received the Nobel Memorial Prize for showing that real-life commons can be, and in fact often are, managed effectively without need for top-down regulation. The resources need to be well defined (e.g., local fishing stocks, shared grazing land, collection

---

*Their proposed solutions included to "limit breeding" and practice "lifeboat ethics" that prioritized space in the lifeboat for the rich over the poor. Hardin also opposed nonwhite immigration and multiculturalism, and expressed eugenic views.

of firewood) and managed by a local community who understand the risk of depleting that resource.

In the case of the planet, however, and humanity's carbon emissions, these conditions are difficult to meet. When the commons is not well defined (which it isn't in this case, because the "commons" essentially includes this entire planet) and is not managed by a tight-knit community that understands what it has to lose by mismanaging its resources (which it isn't, because the "community" in this case includes everyone), its sustainable management often requires formal regulations.

WHY WE LACK INCENTIVE TO ACT

Shouldn't our planet's atmosphere, its fresh waters, its oceans, and its land surface, the resources we can't exist without, be protected? Most of us would say, "Yes," validating Ostrom's findings. Then how did the simple issues of environmental protection, pollution prevention, and climate change become so polarized?

It's undeniable that the most rapid growth in income in the history of human civilization was brought about when the combustion of fossil fuels replaced human and animal labor. But that has led to the myth that we must choose between the environment and the economy: as if an economy could float around in outer space without the resources the Earth provides. This conflict is obvious in the U.S. where distrust of government, regulation, and taxes has been burned into the collective cultural psyche since the Revolutionary War and respect for the free market has been elevated by many to the status of a doctrine.

In other countries where we have similarly benefitted from the consumption of fossil fuels and the destruction of our global commons, we might not distrust our governments as profoundly as so many do in the U.S.* However, we value our comfortable lives and are concerned that

---

*In fact, recent polls indicate that Americans' distrust in government continues at historic highs, and that many trust businesses more than they trust the government. But with their short-term perspective on quarterly returns, many businesses aren't best equipped to manage the global commons.

government intervention may threaten them. We may feel as though those advocating for better management of our global commons want to rip the carpet out from under our feet, leaving us cold and uncomfortable for the hypothetical benefit of people we don't know and aren't connected with. We view solutions as posing an imminent threat, to our quality of life or our jobs or the economy. Why? Because as humans, again, we tend to value things more if we fear they will be taken away.

Individually, as the tragedy of the commons describes, we don't perceive an incentive to act. I say *perceive*, because of course there is a great deal of incentive to act; we just don't see it. *The Loss of Nature and the Rise of Pandemics* is a prescient report by the World Wide Fund for Nature (WWF). It was published in March 2020 as the coronavirus pandemic sped around the world. It points the finger at humanity's overexploitation of nature, including habitat biodiversity loss, as one of the factors behind zoonoses and the spread of new diseases. Few people realize that everyday pollution of air, water, or soil is responsible for one in six premature deaths worldwide. And when disaster strikes, who pays to rescue families from the roofs of flooded buildings, replace washed-out bridges, compensate farmers for crop losses, and provide other forms of disaster relief? Each one of us does, through our taxes and our rising insurance premiums, as well as the costs we face individually for Sheetrock after a hurricane, trucking in water and buying hay bales to feed cattle during a drought, and the devaluation of our home if it's located in a high-risk area for flooding.

But because we're not paying *directly* for our personal exploitation of the environment, our pollution of the air, water, and soil, the accumulation of millions of tons of plastic waste in the oceans each year, and our heat-trapping gas emissions, we each individually lack the incentive to reduce our impact on the global commons that is the Earth. So is it any surprise that there is such a rancorous, ideological reaction to solutions that, by definition, require collective action? If we can't see the risks clearly and up close, why *would* we act?

WHY IT'S NOT JUST A PEOPLE PROBLEM

Almost every time I talk about the challenge of managing our global commons, there's at least one person who says, "Well, there's an obvious solution—population control." And this idea is nothing new. Enforced limits to "breeding," as Lloyd called it, were one of the objectional conclusions both he and Hardin drew from their work. But, as Betsy Hartmann points out in her classic *Reproductive Rights and Wrongs: The Global Politics of Population Control*, "high birth rates are often a distress signal that people's survival is in danger." It's poverty and patriarchy that are responsible for high birth rates, she argues, not the other way around. As women's status improves, birth rates fall.

So while it's tempting for male theorists in rich countries with low birth rates to lean back in the armchair of life and opine on such issues, the reality of a woman's life, particularly in low income countries with the highest population growth, is very different. It's not about giving women fewer choices; it's giving them more. The approach of enforcing control also ignores the fact that global resources are not used in a manner that is either equal or fair. "Overconsumption by the rich has far more to do with climate change than population growth of the poor," Hartmann explains. "The countries where birth rates remain relatively high have among them the lowest carbon emissions per capita on the planet."

To quantify inequalities in the distribution and use of the Earth's resources, in 1998 Swiss sustainability expert Mathis Wackernagel and Canadian urban planning researcher William Rees developed a measure called the *ecological footprint*. It can be calculated for countries or individuals. The footprint defines the resources needed to support one person, measured in "global hectares." These are units of equivalent global biocapacity, roughly the amount of average-quality land each person needs to provide all the food, energy, and other materials that they use.

People in Canada and the U.S., for example, have an average ecological footprint of around 8 global hectares. The average Australian is a 7, the average Brit is a 4. I'm pretty frugal for a North American but even mine is a 5, according to the Global Footprint Network's calculator. In China, they average 3.7; in India, 1.2; and people in countries like Pakistan, Mozambique, and Malawi, less than 1. If everyone in the world lived like the average North American, we would need five planet Earths to support them. As it is, we are already running at a global deficit of 1.1 global hectares per person, using up resources faster than they can be replaced. That's the very definition of "unsustainable."

When we look at carbon emissions, the discrepancy is even more stark. The average American emits about 16 metric tons of carbon dioxide into the Earth's atmosphere annually. Australians produce even more, 17 tons per year, while Canadians produce about the same, 16 tons per year. This represents about four times the global average. It's the amount of carbon dioxide that would be emitted by driving a midsized car one and a half times around the circumference of the planet (60,000 km or nearly 37,000 miles, for reference) once each calendar year. It's the same amount that would be generated by three people living in the U.K., where per capita emissions average 5.5 tons per year, or by twenty-four people in Zimbabwe, and more than forty people living in Yemen today. All told, according to Oxfam, the richest 10 percent of people in the world are responsible for over 50 percent of global emissions. The richest 1 percent produce twice as much carbon as the poorest 50 percent. And while this analysis puts the focus on individuals, in practice the biggest polluters, proportionally, are the big oil and gas corporations, who have a vested interest in encouraging people to continue to burn fossil fuels. And the biggest single institutional consumer of fossil fuels in the world? It's the U.S. military.

LET THEM (STOP) EATING CAKE

In the last chapter I talked about how just one hundred corporations have been responsible for 70 percent of heat-trapping gas emissions since 1988. I also mentioned how, rather than planning for a carbon-free future, some decided to invest in muddying the waters and discrediting the science instead. So you probably won't be surprised to hear that it didn't take long—seven years to be precise—before one of these companies recognized the potential of the ecological footprint concept to shift the blame.

In 2005, the individual *carbon* part of the footprint was extracted and popularized by a British Petroleum (BP) advertising campaign. It included a tool you could use to measure your (not their) carbon footprint, and what a success it was! This concept is now so endemic that I use it myself, to better align how I live with what I believe. But as important as individual choices are, they aren't going to solve this problem. Being a scientist, I've calculated this.

Individual choices control at most 40 percent of emissions in wealthy countries. If you assume that the 28 percent of people in the U.S. who are Alarmed about climate change are willing—and financially able—to cut their carbon footprint in half, that would mean no more than a 6 percent drop in U.S. emissions. Add in everyone who's Concerned, and you can get to 10 percent, maybe. So it's hard to see BP's move as anything but cold-blooded in the extreme: inducing guilt in the public so we'll be too busy blaming ourselves and one another to notice while the richest companies in the world continue to grow their bottom line at the planet's expense.

Thankfully, perspectives are starting to shift. A few years ago, I was checking into my flight at London Heathrow Airport on my way home from one of my bundled trips. I couldn't help overhearing the conversation of the woman checking in beside me. She was explaining to the agent that she worked for BP and had been in London for meetings.

"What's BP?" the woman at the desk asked.

"Well, it used to stand for British Petroleum," the traveler explained, "but they've changed the name because we don't just do petroleum anymore."

And, sure enough, the walls around the airport's security checkpoint were plastered with giant advertisements from BP heralding our solar- and algae-based future, featuring images of children running through green fields under sunny blue skies. In 2020, they became the first major oil and gas company to announce a carbon-neutral goal by 2050. It's a significant sea change, publicly committing to alter their entire business model. I recognize them for it, and I am grateful for all who applied pressure both from the inside and the outside to make it happen. How and even if they'll get there is still anyone's guess, though—and meanwhile, other oil and gas companies continue to employ the blame-and-shame-the-consumer approach today.

In 2019, for example, the CEO of Shell—number three on the richest companies list and number six on the carbon emitters list—told a group of CEOs in London that eating strawberries when they are out of season and buying so many clothes is the problem. "I have three daughters, they are all quite fashion conscious," he said. "I like to point out to them, having something new for every season four times a year is creating quite a significant ecological footprint, have you realised that? Because they are all about climate change."

It's true that the fast fashion and food industries could do with a hard look at their ecological and carbon footprints. But strawberries and clothing? That's what the CEO of the company that has produced over 8,000 million tons of carbon emissions, emissions that it would take 200 billion trees to remove from the atmosphere, says we need to do, to address climate change? And even if he said Shell was going to plant those trees, which he didn't, we'd need more than five times the abandoned agricultural land in the world just to offset *his company's* emissions.

Solving climate change and resource scarcity is not as simple as the analogy of the global commons might make it out to be. Even if the human population were somehow able to be cut in half, say by supervillain Thanos in the Marvel movie *Avengers: Infinity War*, but everyone remaining still lived like the average North American, we'd still need two and a half planets to supply all our needs. If human population were somehow, heaven forbid, reduced to only 10 percent of its current level, but those were the 10 percent richest in the world (as Hardin's "lifeboat ethics" would argue) and the fossil fuel industry continued on their current trajectory, we wouldn't even cut carbon emissions in half. As appealing as "population control" and personal responsibility can be to some, the math just doesn't add up. It's the system we all live in that must change.

THE CLIMATE POTLUCK

"Walking out is not an option. We don't get to give up. This planet is the only home we'll ever have."

MARY ANNAÏSE HEGLAR, *ALL WE CAN SAVE*

"I don't think the U.S. should be part of the Paris Agreement—it's unfair."

AMERICAN MAN TO KATHARINE, SPEAKING OF CHINA

At the Paris climate conference in 2015, it struck me that a climate agreement is like an international potluck dinner. Just as each guest's culture, history, and resources are expressed through the food they bring to share, in the same way each country comes to the negotiating process with its own set of values, goals, resources, and expectations. No one person brings a complete meal, but once all the food is assembled, there is supposed to be enough for everyone.

Growing up, I loved church potlucks. Toronto is one of the most international cities in the world, and our congregation reflected that. Instead of the usual Hayhoe Sunday dinner of well-done beef and roasted vegetables, there would be Egyptian baklava, Jamaican beef patties with turmeric and suet crusts, whole-grain German bread, genuine Italian lasagne, and Caribbean rum cake so alcoholic we had to steal bites when my mom wasn't looking.

In a similar way, in 2015 each country brought their Intended Nationally Determined Contributions (INDCs) to Paris. Before setting their INDCs, they reviewed their emissions and reduction options to determine how much action they could agree to, and of what type. For example, India planned to replace all incandescent lamps with LEDs and accelerate its renewable energy growth; the EU had already capped carbon emissions from heavy industry; and Bhutan was preserving its forests. In addition to calculating carbon sinks and storage, many rich nations also estimated the amount they could contribute to poorer nations suffering the impacts of climate change.

As of 2021, though, there still isn't enough food at the Paris potluck. Some guests have large appetites, while others are near starvation and have little to offer. Current pledges don't match the reductions we need to hold warming below 1.5°C or even 2°C. Only some of the food we need is on the table.

So how do you get more food? A few countries are already bringing enough. Some countries are barely bringing a single serving. And then some countries, which sadly number some of the largest emitters in the world, are bringing nothing to the table at the national scale, according to the Climate Action Tracker. Until 2021 these included the U.S., and they still include Russia, Ukraine, and Saudi Arabia. Yet they're still expecting to eat, because climate action benefits us all, regardless of who implements it. In environmental economics, this is what's known as "free-riding." It's like a wealthy miser who shows up empty-handed to your holiday dinner, then tries to leave with all the leftovers and half your wine rack.

PLANNING FOR THE GLOBAL POTLUCK

This is the last and perhaps most serious implication of the global commons. If enough of the "guests" don't carry through on their promises to bring food, the meal will be sparse and the world will go hungry. If it were really a potluck dinner, you could show people to the door. But you can't show entire countries to the door when we share the same planet. So how do you enforce a global target?

The first resort is often peer pressure and shaming. The biggest of the COP or "Conference of the Parties to the United Nations Framework

Convention on Climate Change" meetings—like the one in Copenhagen in 2009, Paris in 2015, Glasgow in 2021, and beyond—provide ample opportunity to publicly highlight each nation's commitments and capabilities. One country's reductions, which may have seemed ample and sizeable back home, may suddenly shrink when displayed on a global stage side by side with other efforts from similar economies. Others may improve on comparison.

Peer pressure and shaming can be effective, under two conditions that are just as true of countries as they are of individuals. First, does the opinion of those applying the pressure matter to those being pressured? And second, do those being pressured believe there are viable ways for them to do what's being asked of them? Few enjoy being publicly vilified as the person who dragged out the single-serving tart from the back of the freezer while their neighbors brought a fresh-made pie for twelve. Similarly, winners of the Climate Action Network's "Colossal Fossil" award, handed out each day of each COP, probably aren't overjoyed to receive it. But unfortunately, in the case of many of the most recalcitrant countries, the answer to both of those questions above is still largely *no*. Shame has not caused them to deviate from their course.

Another option is to impose economic mechanisms such as border tariffs and sanctions. These can be structured to provide economic incentives to a country to comply with its Paris targets. But they could also run the risk of backfiring economically on the countries imposing them. Having other organizations, such as multinational corporations, exert their influence might help. Perhaps most effectively, though, there's self-interest: recognizing how climate change affects each country, and how much it might actually benefit that country to act.

MOTIVATING COUNTRIES TO CARE

This last option is why I and my colleagues spend so much unpaid time, time that we can ill afford to spare from our own research, on the big national and international assessments such as the IPCC reports and, in the U.S., the National Climate Assessments. These reports draw on thousands of individual scientific studies. Authors meticulously categorize and quantify the impacts of a changing climate on each region and sector, just as our California team did for that original study so long ago. Today, pointing out the difference in impacts between higher versus lower future emissions is de rigueur; the IPCC 1.5°C report released in 2018 took this even further, distinguishing the impacts of 1.5 versus 2°C of warming.

The Fourth U.S. National Climate Assessment was authored by over four hundred federal and academic scientists. It had taken us three years to write. Nearing the end of 2018, we hadn't heard anything from the Trump administration about when or even whether it would be released. The Monday before American Thanksgiving I was at my in-laws' house, in full pie production mode. The counter was covered in apple peels and flour, and so were my hands. My phone chimed. It was a message from a federal colleague. "We've been told the report is to be released *this Friday!*" it said. "Sending proofs right away!"

I immediately reached for my laptop. With floury fingers, I opened the PDFs to see what I had to do and try to figure out how long it would take. The pies went straight into the freezer and the next sixty hours were a flat-out marathon for all of us, with emails and phone calls through the day and well into the night. I finished my own final checks in the car on the way to Thanksgiving dinner and sent off my documents in the next pocket of phone coverage we found driving through rural Virginia.

"Black Friday," the day after American Thanksgiving, is traditionally a "dead" day when it comes to the news, so I was suspicious that the Trump administration might have chosen it on purpose. My suspicions weren't allayed when, in response to Volume 1, White House spokesperson Raj Shah stated, "The climate has changed and is always changing," and when Volume 2 was released, White House spokeswoman Lindsay Walters claimed it was "largely based on the most extreme scenario, which contradicts long-established trends." In reality, as you now know, according to natural cycles the temperature of the Earth should be gradually cooling at this time, not warming. And regarding the likelihood of various scenarios, one of the chapters I led concludes, "The observed increase in global carbon emissions over the past 15–20 years has been consistent with higher scenarios."

The National Climate Assessment's more than two thousand pages exhaustively document how and why climate change matters to every aspect of the U.S. They also show how our choices will determine the future. Even if the federal government wasn't listening, though, others were. When Trump announced he'd be withdrawing the United States from the Paris Agreement (a move reversed by Biden hours after he was sworn in as president in 2021), it spawned the We Are Still In movement. Now known as America Is All In, it has grown to include over two thousand businesses, five hundred cities and counties, twenty-five states, twelve tribes, and many other institutions. Its members represent 65 percent of the U.S. population and are committed to meeting their Paris goals. One recent recruit is the city of Houston, home to many of the U.S.'s largest oil and gas companies. In April 2020, Mayor Sylvester Turner announced the Houston Climate Action Plan, a comprehensive strategy for the city to reach net zero carbon by 2050. It also laid out the city's approach to build resilience to rising seas, stronger storms, and the expected changes in extreme heat and heavy rain that were my own contribution to the plan.

HOW COUNTRIES CAN CAP CARBON

Once a country understands it's in its own best interest to act, policies to manage the global commons can be legislated, implemented, and enforced much more effectively. For each country, sector, or region, there are two main economically efficient policy mechanisms that can be used. The first is known as "cap and trade," the second as "carbon pricing." Both have been studied by economists for decades. Both have already been implemented in different parts of the world. And both offer a way to engage the economy in climate solutions.

Under cap and trade, each company is allocated allowances (or increasingly, buys them at auction) to emit carbon up to a certain limit.

When they reach the limit, they can pay a hefty penalty or buy allowances from other businesses that still have some in hand. If one company is able to reduce its emissions affordably, it might do more than required. It could then benefit financially by selling the remaining "credits" to another company for whom reductions were more costly. In this way, reductions happen where they are cheapest first. It's like one guest at the supper who's not a great cook paying another to bring a contribution on their behalf instead of making it themselves.

In 2005, the European Union set up the world's first international emissions trading program. The cap-and-trade system covers eleven thousand companies in power generation and heavy manufacturing plus the airlines that fly between their countries. These sectors represent around 45 percent of E.U. greenhouse gas emissions. To cut sectoral emissions, the cap (total number of rights to pollute) is reduced every year, and companies are fined heavily if they exceed their allowances.

Although it took a while for the system to start working as planned, by 2020 emissions from sectors covered were estimated to be 21 percent lower than in 2005. Even earlier, a cap-and-trade approach successfully helped reduce North American sulfur dioxide emissions from coal-fired power plants that were causing the acid rain in the 1980s and 1990s. A similar approach has been used to reduce carbon emissions in the U.S. Northeast since 2009, and in California since 2013.

HOW COUNTRIES CAN PRICE CARBON

Cap and trade fixes emission reductions by letting the price of carbon adjust. But it's also possible to fix the price of carbon, and let emissions adjust. And that's why there's a second mechanism that's being applied today around the world: to simply pay for what you eat at the potluck. If someone brings more food than their share, they get to collect a reimbursement for the extra from the pot.

As economist Kate Raworth explains in her book *Doughnut Economics*, classical economists treat the economy, or the "dinner table," as a closed system. They assign no value to either the external resources that power it, including fossil fuels, nor the waste that comes out of it, including the pollution and heat-trapping gases that drive climate change.

To correct these market distortions, over a century ago British economist Arthur Pigou introduced the concept of taxing an "externality," as economists call a cost or benefit that occurs outside the economic system. Carbon taxation is based on the premise that we need to be paying the full cost of using fossil fuels—and we are not. Putting a price on carbon emissions levels the playing field for clean energy, can potentially neutralize subsidies, and can even assign a value to removing carbon from the atmosphere.

In 2018, economist Bill Nordhaus received the Nobel Memorial Prize for applying cost-benefit analysis to calculate the number you'd need to implement a Pigouvian tax on carbon emissions. That number is the social cost of carbon, and it's intended to represent the economic damage from heat-trapping gas emissions, in units of dollars per ton. Today, simply due to existing regulations, the U.S. already effectively operates with a carbon price of $17 per ton. In Canada, which has explicit carbon pricing legislation, carbon is currently priced at $40 Cdn per ton, and will increase by $15 Cdn per year starting in 2022. Globally, the average price is just $2 per ton. Clearly, this number has a long way to go if we want to meet our Paris goals.

Nordhaus's model estimated an appropriate cost of $40 per ton would be enough to prevent dangerous levels of climate change, but that was in 1992. Today, that value is woefully inadequate. As the late economist Marty Weitzman pointed out, such a low cost fails to account for how climate risks can circle back and bite the economy in the rear, so to speak. When those risks are considered, as Weitzman and coauthor Gernot Wagner write in their book *Climate Shock*, there is literally no way to get a carbon price that's less than $100 per ton—and many analyses come up with far higher numbers. "Nordhaus's model implicitly assumes that climate damages are worse when we are richer, and that we should start low and increase the price of carbon over time," Gernot told me. "But what if climate change makes us poorer every step of the way?"

Economists recommend putting a price on carbon that starts low and increases every year until it reaches the true social cost of carbon. This sends a price signal. Instead of paying $1 for a cheap burger at a fast-food restaurant or $3 for a gallon of gas in the U.S., where it is very cheap, we would (eventually) pay the actual price, one that takes into account how much carbon was put out into the atmosphere to create that beef patty or that will be produced when the gallon of gas is burned. If that beef came from Brazil, the price would take into account the Amazon rainforests that are being cut down to make more room for grazing cattle. If it came from North America, the price would reflect our agriculture system, where most cattle eat feed that was grown and processed using equipment powered by fossil fuels. So if we wanted to save money, we might eat less beef and more chicken and vegetables. Similarly, the price for gas at the pump would be higher, so much more that when we did the math, a used electric car might save us money compared to the gas-guzzler we currently drive.

A key component of carbon pricing is how the funds are used. From an ethical perspective it is essential to ensure lower- and middle-income families who spend a greater proportion of their income on food and gas and bills are not harmed by the carbon price. Some of the income can also be used to address environmental justice concerns and invest in and accelerate efficiency improvements, public transportation, and other carbon reduction strategies.

DO THESE POLICIES WORK?

A comprehensive analysis comparing the CO_2 emissions of forty-three countries that have some sort of carbon price at the national or subnational level with ninety-nine countries that don't have one showed that carbon pricing slowed emissions' average annual growth rate by about 2 percent. For each dollar increase in the cost of carbon, the country's emissions growth rate decreased by about 0.25 percent.

The Canadian province of British Columbia introduced a price on carbon in 2008, opting to return all the proceeds to taxpayers. Fossil fuel consumption decreased by more than 17 percent in just five years. Most of the decrease came from efficiency improvements, with some contribution from increases in clean energy. Personal provincial income tax rates dropped to the lowest in Canada, corporate tax rates were some of the lowest in North America, and BC's economy slightly outperformed the rest of the country during that time. From 2007 to 2019, Alberta priced the carbon produced by large emitters. In 2017, this price was extended to the entire economy in the form of sales taxes. All lower-income and many middle-income households, about 60 percent in total, received direct tax rebates to offset their increased costs of living. While I was there in 2018, I spoke to some of the government workers whose job it was to travel around the province to talk to people about the new carbon price. Many communities whose economy was based on the oil fields were hostile, they told me; until they learned that some of the revenues would be used to help people retrofit their houses to increase their energy efficiency and save on fuel bills. Then, enthusiasm was sky-high. People could see the individual benefit to themselves: the carrot, not just the stick. By the time a federal price on carbon was put in place across Canada by Prime Minister Justin Trudeau in 2019, there were four provinces with a price on carbon—BC, Alberta, Ontario, and Quebec—and these four led the country in economic growth.

As of 2020, according to the World Bank, sixty-four carbon pricing initiatives have been implemented in forty-six countries worldwide. These represent a total of 22 percent of global greenhouse gas emissions. There is surprisingly bipartisan support for this approach in the U.S. The Climate Leadership Council bills itself as "an international policy institute founded in collaboration with a Who's Who of business, opinion and environmental leaders to promote a carbon dividends framework as the most cost-effective, equitable and politically-viable climate solution." Founding organizations include ExxonMobil, Chevron, BP, and Shell (yes, you read that right), as well as AT&T, Microsoft, and Santander. For the U.S., their carbon pricing plan aims to cut carbon emissions by 57 percent by 2035—consistent with the 1.5°C Paris target—while creating 1.6 million jobs. Most people would agree that sounds like a pretty good deal.

Here's the problem, though: unless every country participates, legislation that cuts fossil fuel demand, and therefore lowers prices, can indirectly encourage nonregulated countries to up their consumption. German economist Hans-Werner Sinn calls this "the green paradox," and it explains, in part, why global carbon emissions continue to climb even as more and more climate policies are put into place. Thanks in great part to its health impacts, China's coal consumption largely plateaued after its rapid growth in the early 2000s. But China's coal production continues to grow. So at the same time as it's investing trillions in wind, solar, and even long-term technologies like nuclear fusion at home, China is also building hundreds of coal-fired power plants in other countries, like Pakistan and Vietnam, so it can sell them its coal.

That's why our potluck really has to be global. If not, it won't succeed.

EVERYONE NEEDS ENERGY

"Energy managed wisely gives us health and wealth; managed unwisely, it makes us sick and poor."

MICHAEL WEBBER, *POWER TRIP*

"They're installing my wind turbines this week. Me and my neighbor Mattie, we're going to take our folding stools and our lunches out and watch them do it. I'm so excited!"

THIRD-GENERATION TEXAS LANDOWNER TO KATHARINE AFTER A TALK

Energy poverty is real—770 million people, representing 10 percent of the world's population, lacked access to any form of electricity in 2019. Another 2.6 billion cook their meals on open fires or lack access to clean cooking fuels. So when scientists caution that 75 percent of global warming is being caused by fossil fuel combustion, and the only way to stabilize climate is to reach net-zero carbon emissions, a common response is that fossil fuels are a moral necessity. "We need them for energy here," I often hear people in North America say, "as well as to help poor nations develop like we did. Getting rid of fossil fuels will *increase* suffering, not decrease it!"

Gladys lives in a small community called Ziossa, at the end of a long dirt road in the Dodoma region of Tanzania. Like many in her community, she is a subsistence farmer who grows maize and peanuts to support her family. Life in sub-Saharan Africa is already challenging. There are nearly 600 million people there with no access to electricity, and women often bear the greatest part of the burden this puts on their household: gathering fuel and water, and doing all their domestic and agricultural work without the labor-saving devices we in rich countries take for granted. Gladys is a widow with six children; after she lost her husband, life got even harder.

Solar Sister is a nonprofit whose goal is to support rural women in creating their own clean energy businesses. When they visited Gladys's community in 2016, she signed up immediately. She'd never heard of solar before, she said, but now she sells solar lights across the region: mostly to women, but some men, too. Solar lanterns quickly pay for themselves as they don't need to be powered by batteries. And they replace kerosene lamps, which can spark fires and generate indoor air pollution that can cause respiratory disease. This business helps Gladys support her growing family, which now includes seven grandchildren, and Gladys is only one of ninety-four entrepreneurs in her region. Together they have reached 1.2 million people across Tanzania with clean energy, based on a business model that empowers women.

In her TED talk *How Empowering Women and Girls Can Help Stop Global Warming*, Katharine Wilkinson explains that "women are the primary farmers of the world. They produce 60 to 80 percent of food in lower-income countries, often operating on fewer than five acres. . . . Compared with men, women smallholders have less access to resources, including land rights, credit and capital, training, tools and technology. Close those gaps, and farm yields rise by 20 to 30 percent." As Gladys found, clean energy opens even more new opportunities.

On the opposite side of the world, northern Canada's plentiful mining resources often lie near or on the traditional lands of First Nations peoples. In such remote areas, diesel is the primary fuel for both towns and mines. But it's expensive and runs the risk of dangerous spills that can contaminate land and even water supply. Wiigwaasaatig Energy is a partnership between mining corporation AurCrest and three local Ojibway Nations in northern Ontario. They're currently purchasing off-grid mobile wind- and solar-generation units and are ultimately hoping to provide forty megawatts of electricity, enough to power over six thousand homes. The Nations own 51 percent of the project and the company owns 49 percent, giving locals the opportunity to benefit from the mine and generate local power at the same time.

WHY ENERGY—NOT FOSSIL FUEL—IS A MORAL NECESSITY

Electricity from fossil fuels has played a key role in relieving poverty and spurring economic growth in many places. But thanks to clean energy advances, we can now achieve the same goals without ruining the environment and our health. So while electricity *is* a moral necessity—fossil fuels *aren't*.

One reason fossil fuels are no longer the future is the simple fact that it's the 2020s, not the 1820s. Most countries in Western Europe, North America, and Australasia developed at a time when coal was the main source of fuel. But given how far we've now progressed toward clean energy, it's not only patronizing but frankly colonialist to presume that everyone else has to use coal, too. It implicitly says to other countries, "You aren't ready for modern cars and cellphones yet. At your stage of development, you get party line telephones and Model T Fords. Check back with us in fifty years."

Further, while many rich countries have enough fossil fuels to supply their needs for decades to come, most developing countries don't have abundant reserves. Africa, where many of those without access to plentiful energy live, has only 7.5 percent of the world's known oil reserves, 7.1 percent of its gas reserves, and 1.3 percent of its coal reserves. Latin America has just 19.5 percent of the world's known oil reserves—and most of it is located in Venezuela and Brazil, where corruption related to the fossil fuel industry runs rampant and little of the proceeds go to those most in need. It's the developed and wealthy nations of North America, Europe, the former Soviet Union, and the Middle East that have the lion's share, 70 percent of known oil reserves, 79 percent of gas, and 56 percent of coal. So expecting poor countries to develop in exactly the same way as rich nations isn't what I would consider to be moral. Quite the opposite: instead of enabling them to stand on their own feet and supply their own energy, it's inviting them to a lifetime of indebtedness to the rich countries who want to sell them fuel.

Of course, no country needs anyone else's permission to make their energy choices. Today, many developing countries can leapfrog over obsolete technologies to newer, cleaner forms of energy, just like they've already done with cellphone technology. In the past few years, developing countries installed more clean energy than fossil fuel-powered electricity generation due both to increasing demand for electricity in general and the plummeting costs of solar and wind. By 2025, renewables such as solar and wind will generate more electricity globally than coal. According to Bloomberg, the top five emerging markets for low-carbon energy sources are India, Chile, Brazil, China, and Kenya. "It's been quite a turnaround," said Dario Traum, a senior associate at Bloomberg New Energy Finance. "Just a few years ago, some argued that less-developed nations could not, or even should not, expand power generation with zero-carbon sources because these were too expensive. Today, these countries are leading the charge when it comes to deployment, investment, policy innovation, and cost reductions."

In 2019 over 70 percent of new electricity installed around the world was clean energy. That number soared to over 90 percent in 2020 during the pandemic, and it's changing people's lives. Children can study for school in the evening. Women can walk around more safely. Cottage industries offering to charge people's cellphones or pump water are springing up.

Yes, we need energy. Energy is one of the primary ways that we can address and tackle poverty, lack of access to clean water, and sufficient food, among other development goals. But today, for the first time in hundreds of years, fossil fuels don't have to be the source of it. Instead, as Norwegian psychologist and economist Per Espen Stoknes says, "climate change is an opportunity for economic development—an entire energy system has to be redesigned from the wastefulness of the previous century to a much smarter mode of doing things. It's a great opportunity to improve global collaboration and knowledge sharing and to create a more just society."

CLEANING UP OUR ELECTRICITY

Decarbonizing the electricity sector is a low-hanging fruit. That's where about 25 percent of our carbon emissions come from globally, and it's also where change is happening fastest—not just in developing countries. Almost 23 percent of the electricity on the Texas grid in 2020 was generated by wind, eclipsing coal for the first time. Each time I drive out of the city, new wind farms are springing up—with longer and longer blades—on the flat, windy landscape of the High Plains. Knowing that Glasgow would host the 2021 global climate conference inspired Scotland to hit 54 percent clean energy in 2016, 76 percent in 2017, and 97 percent in 2020. Other countries at or near 100 percent clean energy include Iceland, Norway, Paraguay, Costa Rica, and Uruguay.

The reason we're seeing wind farms spread across America's heartland and new solar installations being built across the Southwest—and Mexico, India, Morocco, the United Arab Emirates, and elsewhere—isn't because of government subsidies. It's because prices for renewable energy have dropped so low that subsidies are no longer needed to support them. They can even compete with highly subsidized fossil fuels. From 2010 to 2020, the price of building new wind power dropped 50 percent, and solar costs fell below those of fossil fuels in many places—as little as 3 cents per kilowatt hour in countries like Mexico, Peru, India, and Dubai. In 2010 solar and wind accounted for 4 percent of total global electricity capacity; by 2019 they made up 18 percent, representing a $2.6 trillion investment.

Hang on, you might be thinking. Did you just refer to fossil fuels as "highly subsidized"? Yes, I did. According to the International Monetary Fund (IMF), fossil fuel use is subsidized to the tune of 6.5 percent of global GDP, or nearly $165,000 USD per second. Nearly half of that goes to coal, then petroleum; only 10 percent to natural gas. In the U.S., the IMF estimates that fossil fuel subsidies top $600 billion per year. That is slightly more than the Pentagon's budget, ten times what the U.S. spends on education every year, and more than twenty times the clean energy subsidies. What form do these subsidies take? Some are the result of tax breaks, direct production subsidies, and leases on public land at far below market rates. Other subsidies come in the form of costs on people and land, called *negative externalities* by economists: air pollution, asthma, cancer, land degradation, water contamination, and more, caused by fossil fuel extraction and combustion but paid for by us.

Once you understand just how uneven the energy playing field is, statements about the economic viability of clean energy sound even more transformative. One study finds, for example, that the entire U.S. power grid could be transitioned to 90 percent renewable energy by 2035 at no net cost, and with a reduction in average electricity costs of 13 percent. It would also avert $1.2 trillion in health and environmental damages and 85,000 premature deaths. Globally, the International Renewable Energy Agency's 2019 report on global energy transformation finds that "for every $1 spent for the energy transition, there would be a payoff of between $3 and $7." The report also points out that as renewable energy would require fewer subsidies than fossil fuels, this amounts to a $10 trillion global savings through 2050. For perspective, it's estimated that the U.S. military has spent more than $5 trillion since 2001 fighting the "Global War on Terror."

Manufacturers are increasingly generating their own power on-site from renewable sources, too: popular brand Method Soap's factory outside Chicago features an on-site wind turbine, a green roof, and solar "trees." Big tech firms including Google, Facebook, Amazon, and Microsoft bought more renewable energy than anyone else in 2019, and all have either achieved or committed to achieving 100 percent clean energy.

GOING THE EXTRA ELECTRIC MILE

When we think about cutting carbon emissions, our minds often jump immediately to solar farms and wind turbines. To reach its full potential, though, a clean electricity sector needs to be paired with two other strategies: good old-fashioned efficiency, because the cheapest form of energy is the energy that you don't use, and cutting-edge technology that electrifies cars and trucks, home heating, and industrial processes currently powered by other types of fossil fuels.

Efficiency isn't only about energy-saving lightbulbs, or efficient appliances, or turning off your computer. These do matter, and options are much more attractive than they were twenty years ago. That's when my dad first got the efficiency bug and swapped out all our regular lightbulbs with old-school compact fluorescents. They took two minutes to light up and gave everyone in the room a gray complexion, as if we'd just been rolled out of a drawer at the morgue.

LEDs today come in all shapes and colors, and their costs have dropped so low that it's cheaper now to buy an LED than to pay for the electricity a free incandescent bulb would use. But when we talk about large-scale efficiency, there's a lot more to it: more efficient cars and vehicles, retrofitting buildings, smart home technology, industrial efficiency, computerized optimization of freight transport and airlines, and the efficiency of the entire electrical grid.

Globally, Germany and Italy are in top place on the annual International Energy Efficiency Scorecard. Canada and the U.S. are tied at number ten on the list. The good news, though, is that energy efficiency improvements—across the industrial, transportation, and building sectors—could actually cut U.S. carbon emissions in half by 2050. And energy wasted is money lost, so efficiency improvements usually have a short turnaround time before the initial investment starts saving money.

The even better news, though, comes when you add in the second half: electrification. There, physicist Saul Griffith has a plan. It's called "Rewiring America" and in it, he calls for tripling electricity production and electrifying everything we can, as soon as possible. This includes heating, industry, and most of all, transportation.

The biggest chunk of global transportation emissions, almost 40 percent, comes from passenger cars and SUVs. These are changing as the cost of electric cars drops and more and more companies get on board. Volvo announced recently that as a company it is aiming to become carbon neutral by 2040 and is planning a fleet of electric and hybrid vehicles. In a 2020 Super Bowl ad with basketball star LeBron James, General Motors proclaimed that its infamous gas guzzler, the Hummer, would be reborn as an electric 1,000-horsepower pickup truck in 2021. And in 2021, GM announced it would sell only zero-emission vehicles by 2035.

As of 2020, twenty countries—ranging from rich ones such as Norway and Sweden to low income ones such as India and Sri Lanka—have announced they'll be banning sales of new gas or diesel cars at some point in the future, with estimated dates ranging from 2025 for Norway to 2050 for Costa Rica. China has announced it will be doing this, too, although with no date set as yet. California, the fifth largest economy in the world, plans to ban the sale of new fossil fuel cars by 2035. This might sound far away, but I saw our first electric city bus here in Lubbock in 2020. If we've got them in the second most conservative city in America, the world really must be changing.

What will electrification accomplish? According to Saul, it's a big part of ensuring the U.S. meets its Paris goals, and it will also create 25 million new jobs. "Will our lives change?" he asks rhetorically. "The surprising answer is, not a lot. Those things that will change, though, will be for the better: cleaner air, cleaner water, better health, cheaper energy, and a more robust grid."

REACHING FOR THE HIGHER ENERGY FRUIT

Electricity generation is a no-brainer, but what about other sources of carbon? Industry is responsible for 21 percent of emissions worldwide. These come from construction, manufacturing, mining, and from chemical reactions when creating materials, particularly cement. Certain heavy manufacturing industries, including those that make glass, steel, and concrete, require extremely high temperatures that, so far, renewable sources have not been able to achieve. But in 2019 a new start-up called Heliogen claimed it had created a way to generate heat between 1,000 and 1,500°C using concentrated solar power—effectively a solar "oven." A Canadian company called CarbonCure is commercializing a method that injects CO_2 into cement as it's hardening into concrete, turning the process into a net sink rather than an emissions source. Technologies like these aren't operational yet, but they show the potential for new development in many areas, including heavy industry.

A big part of a grid transition to clean energy is ensuring a stable supply of power when the sun isn't shining and the wind isn't blowing. Low-carbon-baseload power can come from geothermal, or from nuclear, or from using batteries or other methods to store energy generated by wind and solar. None of these solutions are perfect. Geothermal energy does release some carbon dioxide, although significantly less than fossil fuel–powered electricity generation. For example, Iceland produces about 0.16 million tons of CO_2 from its geothermal energy every year, while the U.S. produces almost 2,000 million tons of CO_2 each year generating electricity from fossil fuels. The mining, extraction, and processing of the materials and rare earth metals needed for wind turbines, solar panels, batteries, and nuclear also have a carbon footprint that, while significantly lower than fossil fuels, isn't zero. Nuclear power has other challenges, from the astronomical cost of building a conventional plant* to the ethics and logistics of extracting its fuel and disposing of its toxic waste. Lingering memories of nuclear disasters and concerns about the proliferation of nuclear weapons add to its complexity. But heartening progress is being made in unexpected places.

First, the price of lithium-ion batteries fell 86 percent from 2010 to 2019 and is expected to continue this slide. In September 2019, the city of Los Angeles approved a record-breaking deal involving solar plus battery storage that would supply electricity at 3.3 cents per kilowatt-hour, and battery costs continue to drop. In terms of mining rare earth metals, the DeGrussa copper and gold mine in the Australian bush, formerly run on diesel generators, is now powered by one of the largest off-grid

---

*Cost is the main reason why the U.S. hasn't successfully built a new conventional nuclear plant in thirty years. One multi-year attempt in South Carolina was finally abandoned in 2019 after spending, as one headline put it, "$9 billion to dig a hole in the ground and then fill it back in."

solar farms on the planet, saving 12 million tons of carbon emissions each year. Efforts to recycle batteries, as Tesla does, and solar panels and wind turbine blades as well, as they come to the end of their lives, are essential to minimizing the costs and emissions of resource extraction. And then there's technology: distributed and smart grids can help move the electricity around in ways that increase efficiency and minimize the need for storage. And when electricity is plentiful, the excess can be used to pump water up to a higher hydropower reservoir; when energy is needed, the water can run downhill to power the turbines.

In terms of alternative energy sources, Iceland is already entirely powered by geothermal energy, and others are getting on board. When I visited Ball State University in Indiana a few years ago, everyone wanted to tell me about their new geothermal energy system. It's the largest ground-source closed system in the country, with over thirty-six hundred boreholes that together replaced four old coal-fired boilers. Most of the emissions in Juneau, Alaska, are generated by the millions of cruise ship visitors that power their economy, so residents have designed and set up their own carbon offset program. Visitors pay to offset the carbon emissions of their trip by replacing local heating oil use with zero-emission heat pumps.

New developments in modular "micro-nuclear"—small reactors that can be transported by trucks and grouped together to make a plant—have the potential to substantially reduce the price of nuclear fission and increase its flexibility and safety. The U.S. Department of Energy is currently investing in construction of a modular nuclear plant at Idaho National Laboratory to be integrated into the Utah Municipal Power System to replace fossil fuels they're phasing out. It's also investing in TerraPower, a fast reactor technology that reduces nuclear waste, a partnership between Bill Gates and GE Hitachi. In the U.K., Rolls-Royce is planning to build fifteen mini-reactors in the next decade. Research on nuclear fusion—which mimics the heat-generation process at work in the Sun and does not produce nuclear waste—is ongoing, with the International Thermonuclear Experimental Reactor being built in France as a partnership between thirty-five different countries. China turned on its own new experimental fusion reactor in December 2020. But while safe, affordable, waste-free, and widespread nuclear fusion may be a part of our long-term future, it's not likely to happen at scale anytime this century—and despite recent technological advances, nuclear fission isn't the all-encompassing antidote its enthusiasts often make it out to be.

EXCITED ABOUT THE FUTURE

Is it really possible to move to a world of net-zero electricity? It won't be easy, but the answer is, emphatically, *yes*. Much of the technology is already in place to get there; what's lacking now is the will and the investment. As businesses and new markets alike find opportunities in the new green economy, it's time for us to move on, with gratitude, from the old ways of getting energy that have served us for so many centuries.

After the Christmas nutcracker talk, one ladder-thin, straight-backed, white-haired woman said it best. "I'm a third-generation landowner," she said enthusiastically, "and they're installing my wind turbines this week. Me and my neighbor Mattie, we're going to take our folding stools and our lunches out and watch them do it. I'm so excited!"

"We all should be," I said. "Because it isn't just about reducing the damages from fossil fuels: it's about the hope for a better future, too."

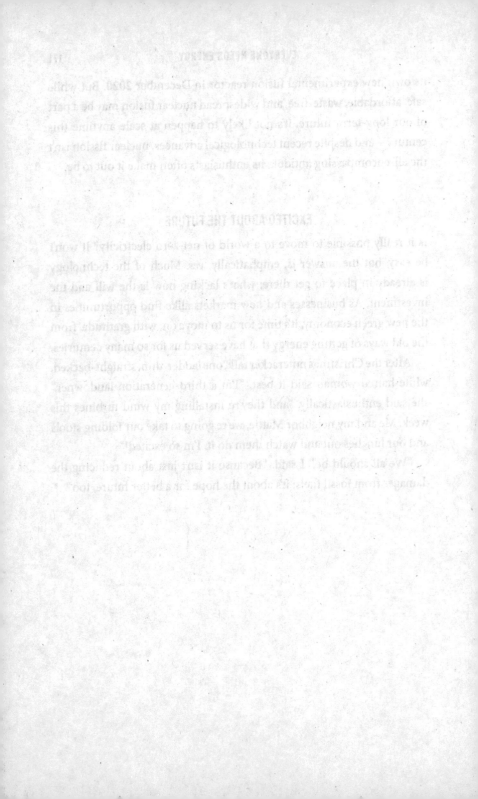

CLEANING UP OUR ACT

"When we plant trees, we plant the seeds of peace and seeds of hope. We also secure the future for our children."

WANGARI MAATHAI, GREEN BELT MOVEMENT FOUNDER AND NOBEL PEACE PRIZE LAUREATE

"Farmers and rural Americans, that's who's going to solve this."

MATT RUSSELL, IOWA FARMER, TO KATHARINE

"I'm leaving climate science," Chris Anderson told me over the phone. Chris and I had worked together on many projects, most recently a set of guidelines for the U.S. Federal Highway Administration to prepare for a changing climate. We're both interested in practical work that provides the information people need to make decisions in the real world.

"Leaving?" I asked in disbelief, finding it hard to understand why someone like him, successful and smart, would be giving up climate science at a time when we need every hand on deck. "To do what?"

"To work for a new biofuel company," he said.

Biofuels . . . I'd heard of those. They were promoted by the George W. Bush administration to pay farmers to grow corn to turn into ethanol. The fossil fuels used to grow the corn and convert it into fuel exceeded any carbon emission reductions from burning it, and the program was only successful because it was heavily subsidized by the federal government. Plus, turning food into fuel? Even kids see the problem with that!

My colleague Michael Webber is an energy expert. In his book *Power Trip* he recounts how, one night, he was watching a documentary on corn

ethanol. He didn't think his eight-year-old daughter was paying attention. But at the end of the movie, she bolted up, grabbed a pad of paper,
and scribbled out a two-page essay called "Why We Can't Use Corn!" that
began with the line, "We should not use corn to make oil because we eat it!"

"That's not what we're doing," Chris assured me when I expressed
my doubts. "And I think this could really make a difference with climate
change. Why don't you come for a visit next time you're in Iowa and I'll
show it to you?" So I did.

The headquarters for Renewable Energy Group, Inc. (REG) lies at the
end of a business park in Ames, Iowa. The company is professional, organized, and clearly influenced by their engineers and scientists; much of
their wall space is covered in infographics and diagrams. It takes agricultural waste and used cooking oil from restaurants and other food producers and transforms them into carbon-neutral bio-based diesel fuel that
can be used in bus and truck fleets with no engine modifications. No new
products and certainly no crops grown specifically for REG go into this
fuel; it's all made from waste products that would otherwise be thrown out.

REG's engineers know that, long-term, many will transition to electric
buses and trucks. For now, though, there often isn't enough infrastructure to
support these types of electric vehicles. A lot of investment is needed for this
solution to become a reality. Today, most buses and other long-haul vehicles
still have internal combustion engines. Compared with using petroleum-
based diesel, REG's biomass-based diesel fuel could cut emissions over the
whole production cycle by 86 percent. This is a great option until a city or a
trucking company is ready to replace their fleet with electric vehicles.

"What happens if you run yourselves out of a job?" I asked.

"We can move on to aircraft fuel, then," one of the engineers replied.
"The world will also be a much cleaner place."

CRACKING THE TOUGH ENERGY NUTS

Sectors that require energy users to carry their liquid fuels with them, like
shipping and aviation, are two of the toughest nuts to crack when it comes
to carbon-free energy. Today, shipping makes up about 2 to 3 percent of

the human impact on climate change. Aviation is responsible for an estimated 3 percent as well. Together, if these two industries were a country, they'd be the fourth largest emitter in the world, between India and Russia.

The International Maritime Organization has committed to reducing shipping emissions at least 50 percent by 2050. Ships such as ferries that make short trips can be electrified. After Norway banned all non-electric cruise ships and ferries from its UNESCO World Heritage fjords, cruise lines immediately began building electric ships, and the first one set sail in 2019. Two other start-ups are looking to bring the past into the future, designing sail systems that could quickly retrofit existing cargo ships with masts and large, rectangular sails they estimate would reduce their fuel use by up to 20 percent. Both the Wind Challenger Project based in Japan and the U.K.-based Smart Green Shipping Alliance hope to have demonstration ships ready by 2021 and are also working on entirely wind-powered ship designs.

With aviation, only one-third of its warming effect on climate comes from burning jet fuel. The remainder comes primarily from the condensation trails or contrails planes leave in their wake, as water vapor condenses around particulates in the exhaust. About a third of aviation emissions come from short-haul flights, and there is significant potential to electrify these. The Israeli firm Eviation has created a prototype nine-passenger electric plane with a 650-mile range that is expected to take its first flight in 2021. Cape Airlines, which serves New England, has already placed an order for a number of them. EasyJet, the popular European budget airline, is partnering with a start-up appropriately known as Wright Electric to develop electric short-haul planes over the next decade.

Electrification doesn't work for long-haul flights. It's simply because of weight. Per kilogram, jet fuel contains about fifty times more energy than current battery technology. For example, the Airbus A380, until production ceased in 2019, was the largest passenger plane in the world. Aeronautical engineer Duncan Walker calculated that while it had a fossil fuel–powered range of fifteen thousand kilometers, with batteries, that range shrank to only one thousand kilometers. Instead, international airlines, under the United Nation's Carbon Offsetting and Reduction Scheme for International Aviation, are looking to reach their goal of any growth after 2020 being carbon neutral via efficiency improvements; alternative fuels made of "green molecules" that don't produce any new carbon when burned, including ammonia, hydrogen, and biofuels; and offset programs, where they pay for carbon emissions to be reduced or for carbon to be removed from the atmosphere elsewhere.

Unlike ammonia or hydrogen fuel, which have to be stored under pressure, biofuels can be swapped out with fossil fuels, no new engines or retrofits needed. What's used to create the biofuel is the challenge, though. It can be synthetic, or it can be made from something that removed the carbon from the atmosphere recently and did not, unlike corn-based ethanol, incur any additional carbon in doing so. Candidates include algae, agricultural waste, and cooking oil; even garbage, manure, and grass clippings could be turned into fuel. In 2019, a team of chemical engineers from University College London won British Airways' Sustainable Aviation Fuels Academic Challenge with a plan to turn household waste into jet fuel capable of powering a five-hour long-haul flight without any carbon emissions. At this point, biofuel is more a question of supply, logistics, and cost rather than technology— another place where a price on carbon would accelerate the transition.

The post-COVID green recovery is lending extra impetus to these plans. To receive a government bailout, France and the Netherlands required Air France and KLM respectively to cut their carbon emissions per passenger in half, relative to 2005, by 2030. United Airlines has been refueling its flights out of Los Angeles with biofuel created from agricultural waste since 2016, and five other airports—Bergen and Oslo in Norway, Amsterdam in the Netherlands, Brisbane in Australia, and Stockholm in Sweden—offer biofuel refueling options.

Some sectors will take longer than others to decarbonize, but change is happening. As Air Emirates president Tim Clark said in 2019, "We [in the aviation industry] aren't doing ourselves any favours by chucking billions of tons of carbon into the air. It's got to be dealt with." The only question is whether it will occur quickly enough to avoid dangerous climate change—and so far, it isn't. That's why we need climate policies: to accelerate the transition that is already occurring around the world.

FARMING THE FUTURE

"Farmers and rural Americans, that's who's going to solve this," says Matt Russell. Matt is a farmer in Iowa. His farm, Coyote Run, lies about an hour south of where Chris works in Ames. Matt has experienced firsthand the pressures small landowners are facing as industrial-scale production continues to spread across the landscape. And he thinks climate solutions offer the answer for him and his fellow farmers to stay in business.

Being in Iowa, Matt has some unique leverage. It's the first state in the U.S. to hold its primary caucus each election cycle. In the caucus, registered members of the Republican and Democratic Parties cast their vote for which of the many candidates for president they want to lead their party. Every major news outlet in the country (and many international ones as well) descend on Iowa, and candidates often spend weeks in advance of the caucus traveling up and down the state. So Matt shares his carbon message with every politician who visits Iowa. As a result, nearly every primary candidate, Democrat or Republican, in the past decade has gotten an earful on what farmers can bring to the climate potluck and how that would benefit middle America.

Forestry, land use, and agriculture may not seem like obvious places to look for climate solutions, but they comprise a huge chunk of heat-trapping gas emissions: 24 percent of the global total, to be exact. The biggest part of those emissions comes from livestock and deforestation. Ruminants like cows, sheep, and goats belch out copious amounts of methane, which as I mentioned before is thirty-five times more powerful than carbon dioxide. And as the world's demand for meat and animal products increases, we're seeing more slashing and burning of virgin forest to create space for grazing, which creates even more carbon emissions.

It's easy to see how a carbon price would help level the playing field by making new technologies such as renewable energy, high-temperature industrial processes, and electrification and biofuels for aviation and shipping more affordable compared to fossil fuels. But there's another benefit to a price on carbon as well. Too much carbon in the atmosphere is bad, but carbon in the soil and biosphere is good. So how can we get it there? Plants are the key.

That's why Matt's right when he argues that carbon farming, smart soil management, and sustainable agriculture are essential to any comprehensive climate plan. Not only that, but their payoff is substantial. Conservation agriculture is an approach that minimizes tilling and soil disturbance. It protects soil with cover crops, leaves waste behind after harvest, and uses crop rotations to manage the soil. Integrating livestock can close the cycle, providing a use for crop residue and waste vegetables while producing manure that can be used to compost and serving as part of crop rotations. Project Drawdown estimates that conservation agriculture could sequester a year's worth of the entire world's carbon emissions and save farmers somewhere between $2 and $3 trillion in lifetime operational costs. Another year's worth could be sequestered by protecting indigenous people's rights to manage their land, and a further one to two years' worth of emissions through managed grazing and the integration of trees, pasture, and animal forage. Traditional practices, from fire management to agroforestry systems, not only increase carbon in the soil and biosphere but also protect habitat and biodiversity. It's yet another win-win.

Like me, Matt is motivated in large part by his faith. When he's not farming, he serves as executive director of Iowa Interfaith Power and Light, an organization whose programs include "Faith, Farms and Climate." It brings farmers together in church basements to talk about climate policies that would help them thrive. Farmers take the concept of stewardship seriously, and that's the very value that informs the actions and attitudes Matt embodies and shares.

People in Texas think that way, too. That's what motivated a group of local sixth graders to partner with another one of my Texas Tech colleagues, soil scientist Natasja van Gestel, to create a science project called "Carbon Keepers." The students measured how carbon levels change in response to drought, wildfire, and fertilizer, and studied how farmers could store carbon as organic matter in soil. Then they went out and shared what they learned with local farmers, ranchers, and community groups. In 2020 they won their category of a U.S. Army web-based science, technology, engineering, and mathematics competition. As I've said before: if something like this can happen in Lubbock, Texas, how else might the world change?

PUTTING CARBON BACK

When I visited his lab, Robert Brown, director of the Bioeconomy Institute at Iowa State University, introduced me to one of the most direct ways of putting carbon back in the soil. "This dark gray powder," Robert said, uncapping a test tube and pouring some biochar into my hand, "is like MiracleGro on steroids." It's basically pure carbon, and if you plow it back into marginal or average soils, it is one of the best fertilizers under the sun. It also sequesters that carbon in the soil, instead of in the atmosphere.

He pulled out two photos from the previous July of tomato plants in his backyard.

"I bought these the same day and planted them with the same soil," he said. "The only difference is that I mixed biochar into one of the pots. This is a picture from eight weeks later. Can you guess which one?"

It wasn't hard to tell: one plant had four or five tomatoes on it; the other looked like it had more tomatoes than leaves. All told, he said, the biochar pot produced several times more tomatoes than the pot without. A process called pyrolysis, burning agricultural waste at high temperatures in the absence of oxygen, allows him to distill the carbon local crops pulled out of the atmosphere just a few months earlier into biochar, along with products, including oils, that can be used for other purposes.

This is similar to the approach taken by SymSoil, a company based in California. They've created a new type of biologically active compost they call "Soil Food Web" that increases soil carbon storage by 2.5 metric tons of carbon per acre per year. Working with David Johnson, a researcher from California State University, they're producing a special mix of biochar infused with fungi that increases the carbon storage to 10 tons per acre per year. It reduces the amount of irrigation needed, as well.

Biochar, crop rotations, and regenerative agriculture aren't new technology. Ancient peoples have been doing these things for centuries to enrich their soils. Many Westernized countries have just neglected them in recent decades, blinded by our fascination with big-box farming and industrialization, cheap pesticides and fertilizers, and the "bigger is better" mentality of the Industrial Revolution. Regenerative agriculture practices encourage us to return to the wiser approaches of previous generations. But at the very opposite end of the spectrum there's also innovative, out-of-the-box technology to suck carbon out of the atmosphere and turn it into something that's harmless. Much of this is led by, again, scientists who've stepped out of academia to develop leading-edge solutions, some of which sound like science fiction.

Scientists have recently discovered how to grind up rocks that then react with CO_2, pulling it out of the atmosphere and turning it into . . . more rocks. This might sound counterintuitive, but anything that moves carbon from the atmosphere back into the lithosphere, where fossil fuels come from, is a big help. About 25 percent of the carbon humans produce is taken up by the ocean, and in 2017 graduate student Adam Subhas at Caltech led a team that discovered how, by adding an enzyme, they could speed up those chemical reactions by as much as five hundred times. Then, in 2019, researchers in synthetic biology from the Weizmann Institute of Science in Israel successfully created a strain of E. coli—a type of bacteria that commonly lives in the intestines of healthy people, feeding off sugars and fats—that instead consumes carbon dioxide.

Climeworks is a small Swiss company that's developed and, even more importantly, brought to market what's called "direct-air-capture" technology: the ability to suck carbon dioxide right out of the air. With carbon pricing, they'd be able to turn this into a solid product that could be stored deep underground, essentially removing the carbon from the atmosphere for millions of years. Lacking that incentive, instead they've turned to products they can make from the captured carbon and sell, from stone at their Iceland plant to the carbon they capture in Switzerland and sell to greenhouses and fizzy drink sellers such as Coca-Cola.

There's also the idea that rather than letting the carbon from burning coal or natural gas escape into the atmosphere, we can trap it and store it underground. There's one operational carbon capture plant in the U.S., in Texas. Petra Nova is touted by many politicians as the future of climate solutions. But the carbon dioxide it captures goes straight to the oil and gas industry, to enhance oil recovery from existing wells. The reality is that carbon capture and storage is nearly always a more expensive option than just cutting emissions in the first place, and there's even a risk that such approaches create more CO_2 than they remove.

The true holy grail of carbon capture has already been achieved in 2018 by a Canadian company called Carbon Engineering: experimentally, that is. They combine the carbon dioxide they suck out of the atmosphere with hydrogen from water and turn it back into liquid fuel. When burned, it can be carbon neutral since it's only releasing carbon that was already in the atmosphere. This has the added bonus of creating a fuel that can be used in places where it isn't easy to trade out liquid fuels for batteries: like ships and aircraft. Carbon Engineering's technology is being used to develop 1PointFive, a new, industrial-scale joint venture between Occidental Petroleum and Rusheen Capital Management, to be built in Texas's Permian Basin in 2022. The plant will be able to capture and store one million tons of CO_2 each year. And, with an appropriate price on carbon, more of these projects could soon get off the ground.

SCALING UP NATURE

Taking up carbon isn't necessarily high-tech. Planting trees can take up massive amounts of carbon dioxide. Trees carry all kinds of side benefits: supporting natural ecosystems and biodiversity, filtering water, and cleaning air. A 2019 study claimed that planting a trillion trees—which could be accomplished on the land currently occupied by parks, forests, and abandoned land today—would take up the equivalent of at least a decades' worth of human carbon emissions. It caused a furor, as it made it seem like climate change might be much easier to fix than we'd thought. The YouTuber MrBeast set up #teamtrees with a goal of raising $20 million to plant trees through the Arbor Day Foundation. As of

2020, the team has already planted over 22 million trees. The One Trillion Trees initiative (1t.org), launched at the World Economic Forum at Davos in January 2020, gained support from many world leaders, including then President Trump. A program called Ant Forest allowed users of the Alipay online payment platform in China to fund 100 million trees in that country alone.

Unfortunately, it's not as simple as it sounds. While tree planting is still an excellent solution, it turned out that there were a few errors in the trillion trees calculation. Correcting them shows that the benefits are closer to a year or two of emissions rather than a decade or two. Yes, tree-planting is great, and we should do more of it. We should also protect and restore forests that are in danger of being lost. As The Nature Conservancy argues, this is one of the most cost-effective natural defenses against climate change. Its Africa Forest Carbon Catalyst, for example, aims to avoid or reduce 20 million tons of CO_2 emissions each year; restore or conserve 10 million hectares of African forest; and create 5,000 local jobs. Initiatives like Cities4Forests take a similar holistic approach, partnering large urban centers with "inner forests" in the city itself, "nearby forests" in the surrounding areas, and even "faraway forests" that sequester carbon and protect biodiversity. So far, sixty-three cities have signed up, from Accra, Ghana, to Detroit, Michigan.

On their own, trees don't offer a magic elixir or a get-out-of-the-climate-jail-free card. But when they're incorporated into all of the other nature-based climate solutions I've already talked about, together they could add up to over a third of the reductions needed to meet our global 2030 target.

ENGINEERING THE PLANET

When it comes to climate solutions, though, the elephant in the room is solar radiation management, or SRM. It's the idea of deliberately interfering with the Earth's atmosphere for the specific purpose of altering its energy balance. In essence, engineering our entire home.

One could argue that humans are in fact already geoengineering the planet with our emissions. As far back as scientists can track the geologic history of the Earth, there is no parallel for the sheer volume of carbon we're pouring into the atmosphere every year. However, the terms "geoengineering" and "climate intervention" are usually applied to a situation where we are doing it intentionally.

One way that can be done is to mimic the effect of a large volcanic eruption on the Earth, by injecting particles into the upper atmosphere. This increases the amount of sunlight that's reflected back to space rather than being absorbed by the Earth, which in turn cools the planet. Another way would be to sprinkle cloud condensation nuclei—simple sea salt—in large areas over the ocean, to make marine clouds brighter and more numerous, so they reflect more of the Sun's energy back to space.

Both of these forms of SRM are scalable and adjustable. This makes them cautiously appealing to geoengineering proponents—and very appealing to opponents of climate action and carbon emissions reductions who often appear willing to approve any plan, however untested, if it will let them continue burning fossil fuels. It's similar to the mentality that led people to tout unproven cures to coronavirus while poohpoohing the tried-and-true practice of wearing masks.

It's concerning, though, that some of these methods are relatively affordable and well within the technological capacity of nations who are being disproportionately affected by climate impacts. What if one of these countries decided unilaterally to geoengineer the planet? That would be like conducting a Phase 1 vaccine trial with the entire human race all at once. Scientists have a fair grasp of some of the side effects, but certainly not all of them. And, as I've said already, this *is* the only planet we have.

There's no question that if we continue on our current pathway overconsuming fossil fuels, we may need all our options on the table. Making better choices over the past few decades would have been far more preferable. Given the risks the impacts pose, there may come a point where some would feel that this type of geoengineering could conceivably be justified. But it is clear that taking such a drastic step could leave us to deal with a whole slate of unintended and unanticipated side effects. And while geoengineering might temporarily drop global temperature, it wouldn't do a thing about all the carbon dioxide that is building up in the ocean.

Today, the ocean is 30 percent more acidic than it was a hundred and fifty years ago. Acidification decreases the amount of calcium carbonate in the water, a mineral that is one of the key building blocks for phytoplankton shells. Phytoplankton produce half the oxygen we breathe and form the base of the marine food chain. Clams, mussels, oysters, and coral need calcium carbonate, too. Sea creatures that grow shells can even see them dissolve in acidic conditions. Acidification's impact on marine life threatens food security, livelihoods, and the global economy.

Not only does SRM not help with acidification, but if it were to stop for whatever reason, the particles would clear out of the atmosphere in a matter of months to years—meaning that all the warming they had offset would be abruptly realized. The precipitous rise in temperature would be devastating. The "cure" being removed so quickly would almost certainly be worse than the disease.

Studying geoengineering to determine whether or not it should even be included in our arsenal of options to combat climate change is wise; that's why Harvard and Oxford have research programs dedicated to it. But it should not and cannot be considered as a first line of defense against climate change. We must reduce and eventually eliminate our fossil fuel emissions, as well as suck some of the carbon we've produced out of the atmosphere. Only then can we slow and eventually stabilize our current rate of warming.

That's why it makes all the sense in the world to reduce our emissions as fast as we can, as soon as we can: because, to paraphrase John Holdren, the more we do now, the less we'll have to worry about the future, and the less risky the solutions we'll need.

TIME TO SPEED UP

"The planet will survive. The question is whether we will be here to witness it."

CHRISTIANA FIGUERES AND TOM RIVETT-CARNAC, *THE FUTURE WE CHOOSE*

"The most important thing an individual can do right now is not be such an individual."

BILL McKIBBEN TO KATHARINE, WHILE ON A PANEL TOGETHER

We often picture the challenge of solving climate change as a giant boulder at the bottom of a huge hill. Only a few people are straining their backs to roll it up, and it hasn't budged an inch. But in reality, as you've seen from the last few chapters, that giant boulder is already at the top of the hill. It's starting to gradually roll downhill in the right direction. There are many millions of hands on it, pushing. Each one we add speeds it up a little more.

We don't have all the technology we need to go a hundred percent carbon-neutral tomorrow. But we do have what we need to get at least halfway there, and we know what to do to get the rest of it in place. Tried-and-true policies like cap and trade and carbon pricing, and targeted investment in research areas like liquid fuels and smart grids, will create the markets and help spur the innovation that will get us the remainder of the way there.

At this point, it's not a matter of whether. It's a matter of when. And at our current pace, despite all the enormous progress that's already been made, we aren't moving fast enough—yet.

It's been more than half a century since those scientists first warned a U.S. president of the risks of climate change. Thanks to huge evidence-gathering and modeling efforts by the global scientific community, most countries in the world have pledged to reduce their emissions. Yet according to the Climate Action Tracker, as of 2021 current policies around the world would limit warming to a best-case scenario of just under 3°C, when we need to keep the rise to 1.5°C or at most 2°C to avoid disastrous impacts. Worldwide, replacing coal, oil, and gas is still happening ten times slower than what's needed to meet climate goals. That's not a lot of action for a threat that scientists have known about since the 1800s and have been warning people about for decades. So how can we speed things up?

DIVESTING FROM FOSSIL FUELS

The fossil fuel divestment movement began in 2010. Students in the U.S. urged their universities and institutions to scrub their endowments of fossil fuel investments and transfer these funds to clean energy and community resilience planning. From there, it spread around the world: from faith-based organizations, such as the Church of England, to dozens of cities, such as Copenhagen, Christchurch, Paris, and Sydney, and even entire countries, like Ireland. Leaders like Bishop Desmond Tutu and environmentalist Bill McKibben are strong advocates for divestment, and organizations such as 350.org and Fossil Free bring people together to encourage their institutions to divest. As Bill says, one of the most important things an individual can do right now is "not be such an individual." Working together to make a large-scale improvement in our treatment of the global commons can be incredibly effective, and the divestment movement is a prime example of this.

Don Lieber is a surgical technician who works in the operating room at Memorial Sloan Kettering (MSK) Cancer Center in New York City—a "pass the scalpel guy," he calls himself. As he heard more and more about the health care implications of the climate crisis, he found himself wondering, "Why on earth is the health-care industry not more active in this?"

The American Medical Association and the British Medical Association have explicitly called for fossil fuel divestment as a public health imperative. So Don started a campaign to get MSK to divest its staff pension and retirement plans from fossil fuel holdings. But despite emphasizing the public health time bomb that climate change represents, he received a polite no, couched in terms of the financial responsibilities of the fund to its beneficiaries.

So he collected signatures from doctors, nurses, and support staff asking the hospital for the option of investing their personal retirement portfolios in fossil free index funds. "If MSK doesn't have the spine to commit to fossil fuel divestment," says Don, "they should at least give the staff the option to do so."

Don respects the medical science performed at his workplace. "These are some of the most published cancer surgeons in the world," he says. "I'm not trying to single my hospital out—I want all of the hospital systems to do this." But the health care sector overall has been mostly silent on fossil fuel divestment, he maintains. "All we are asking," Don says, "is that our institutions' investment practices are held to the same ethical parameters that we, as healthcare professionals, pledge to in the Hippocratic Oath: 'First, do no harm.'"

Don's not alone. Sustainable mutual funds are growing in popularity, and many others, from university faculty to business professionals, are starting to ask the hard but necessary questions about where their savings are being invested. Thanks to people like Don, many organizations have already chosen to sell or otherwise rid themselves of fossil fuel investments for ethical or moral reasons, due to the harmful impacts of fossil fuel use.

Even large financial corporations are beginning to include mention of ethical concerns as motivation for divestment. Goldman Sachs, one of the world's largest investment banks, announced in 2019 that it would no longer invest in Arctic oil exploration due to "harsh operating conditions, sea ice, permafrost coverage, and potential impacts to critical natural habitats for endangered species." In 2020, Michael Corbat, the CEO of Citigroup, said banks should "have the courage to walk away" from clients who refuse to reduce their carbon emissions. And as part of the "green recovery" plan from coronavirus, twelve more major cities around the world, from Cape Town to Vancouver and including London, New York, and Los Angeles, announced they were divesting their pension funds from fossil fuels.

As of 2020, over thirteen hundred organizations and nearly sixty thousand individuals, whose assets totaled over $14 trillion, had either already begun to or promised to divest from fossil fuels. Thirty-four percent of the organizations are faith-based, and 30 percent are philanthropic or educational institutions, emphasizing the importance of the ethical argument. However, 12 percent are government, another 12 percent are pension funds, and a small but growing 5 percent are for-profit corporations. That's because it's not only about ethics or reputation anymore; there are increasingly sound financial reasons to divest as well.

KEEPING THEM IN THE GROUND

A significant proportion of fossil fuel reserves need to stay where they are—buried in the ground—to meet the Paris targets. Specifically, up to 80 percent of known coal reserves, 50 percent of gas reserves, and 33 percent of oil reserves will need to remain unburned if the world is to have any hope of meeting the 2°C Paris target.

Companies banking on the financial worth of these resources are increasingly facing the risk of being left with "stranded assets," resources that they can't use or sell. The cumulative value of stranded fossil fuel assets is estimated at up to $4 trillion, if rapid action is taken, and nearly double that if it is not.

There's also the growing risk climate change poses to businesses. On the front lines of skyrocketing payouts for increasingly devastating disasters, the insurance and reinsurance industry has been concerned about this for years. Other financial entities are finally starting to recognize these risks as well. In September 2020, for example, the U.S. Commodity Futures Trading Commission released a report stating that "climate change poses a major risk to the stability of the U.S. financial system and to its ability to sustain the American economy . . . a major concern is what we don't know." Risk management firms are beginning to offer indices that allow companies to assess their vulnerability and risk, such as one by Verisk Maplecroft whose stated goal is to allow its users to "understand the exposure of your operations, supply chains, and investments to climate change–related risks."

Both of these factors—the risk of stranded assets and of climate impacts—are likely playing into announcements like the one made by Larry Fink, the CEO of BlackRock. It's the largest asset manager in the world, with more than $7 trillion in investments. In a January 2020 letter to global CEOs, he announced the firm would drop investments such as coal that pose "a high sustainability-related risk," adding that "the evidence on climate risk is compelling investors to reassess core assumptions about modern finance."

COUNTING THE COSTS OF CLIMATE IMPACTS

The very real financial costs of continued dependence on fossil fuels are growing increasingly clear. Twenty years ago, for example, Chilean engineer Luis Cifuentes and colleagues calculated that, based on the health costs alone—its impacts on illness and death, lost work days, and more—fossil fuel use in many of the world's largest cities was no longer cost-effective. So why are we still using them? Because those who are bearing the costs of dirty, outdated sources of energy are not those who are reaping the profits. This socialization of impacts is yet another irony, considering how often climate solutions are dismissed as "socialist."

At the global scale, Morgan Stanley estimated that climate-related disasters alone cost the world $650 billion over a three-year period ending in 2018. North America shouldered the majority of those costs, $415 billion, or 0.66 percent of the continent's combined gross domestic product. But present-day impacts are dwarfed by the future costs of not acting on climate. Future costs are difficult to estimate, as they must be done by sector, and most estimates are limited to a few sectors. Even these, however, are staggering. One estimate by Australian academics accounted for estimated agricultural losses, sea level rise, and impacts on human health and productivity, but did not include losses from increasingly severe extreme weather events such as hurricanes or wildfires. Still, it put the annual costs of a 2°C warming to the global economy at $5 trillion and a 4°C warming at $23 trillion.

As I've already discussed, the suffering global warming will cause to resources and economies is not going to be parceled out equally either between or within countries. In the U.S., national GDP would drop by 10 percent under a 4°C warming, but more southern states could see larger decreases, of up to 20 percent. Globally, poor countries are already and will continue to be affected the most. While countries like Canada, Japan, and New Zealand could lose 7 to 13 percent of their possible GDP under a 4°C warming, impacts on sub-Saharan Africa and Southeast Asia are expected to be much greater.

To put these numbers in perspective, global economic losses due to the coronavirus pandemic are estimated to reach $22 trillion between 2020 and 2025. Imagine those same losses happening *every other year* due to a changing climate. Then imagine what would happen if the world actually met the targets of the Paris Agreement instead. According to the WHO, Cifuentes and his colleagues were right on track. "In the 15 countries that emit the most greenhouse gas emissions," WHO says, "the health impacts of air pollution are estimated to cost more than 4% of their GDP [whereas] actions to meet the Paris goals [that would eliminate most air pollution from fossil fuels] would cost around 1% of global GDP." That's a substantial amount, but still far less than the cost of air pollution alone that would be incurred by inaction.

And these estimates don't even include the damages due to disappearing ecosystems and entire species going extinct. They only include what economists are able to price. What they can't price might end up costing us even more, in the long run, than what they can.

WHY IT'S TIME TO MOVE ON

The bottom line is this: Humans have been using fossil fuels for a very long time—all the way back to the coal we were burning in the Middle Ages. While coal, oil, and gas have brought us significant benefits, they have done so at the expense of accumulating a substantial climate debt that is now coming due. At some point, it just makes sense to move on.

This is the approach Stephen Heintz, the president of the Rockefeller Fund, took when he announced its decision to divest from fossil fuels in 2014. He said, "John D. Rockefeller, the founder of Standard Oil, moved America out of whale oil and into petroleum, and we are quite convinced that if he were alive today, as an astute businessman looking out to the future, he would be moving out of fossil fuels and investing in clean, renewable energy."

This is the challenge that lies before us, and it's not a small one. In fact, it may well be the biggest fight our civilization has ever faced. In this fight for our future, though, we're not alone. My hand is on the boulder, pushing it down the hill. So, too, are the hands of millions of others: countries, corporations, and organizations big and small. There are individuals, too. Some are leaders many people have heard of, like Jane Fonda or Michael Bloomberg. Others work behind the scenes, like communicator Susan Hassol, who's been helping scientists explain climate change in plain English for over thirty years, or Yvonne Aki-Sawyerr, the mayor of Freetown, Sierra Leone, who's on a mission to plant a million trees in her city by 2022. There's Pastor Mitch and the Evangelical Environmental Network; Farmer Matt and his Iowa climate stewards; the sixth-grade Lubbock soil team; and all the grown-up engineers busy turning waste, garbage, and even excrement into products we can all use.

We just need to get that boulder rolling faster. And that's where you come in.

SECTION 5:

YOU CAN MAKE
A DIFFERENCE

WHY YOU MATTER

"Example, whether good or bad, has a powerful influence."

GEORGE WASHINGTON

"I've crunched the numbers and it's a financial no-brainer to get solar power."

JOHN COOK'S SKEPTICAL FATHER

"So now I know why climate change matters, and what real solutions look like," you may be thinking. "But what does it have to do with me? I'm not a government or a multinational corporation or a famous person, and I don't have any hope of influencing one, either. What am I supposed to do?"

One of the biggest reasons our actions matter is that what we do changes us. And the other big reason is that what we do and say changes others, too.

John Cook has a PhD in cognitive science. He's a scientist at Monash University's Climate Change Communication Research Hub in Melbourne, Australia. He has studied and written extensively on all the issues I talked about earlier: cognitive bias, motivated reasoning, and the backfire effect. He's also created a phone app, a book, and a series of videos called *Cranky Uncle Vs. Climate Change: How to Respond to Climate Science Deniers.* Suffice to say that when it comes to science denial, he is *the* expert.

But John's also a human, just like the rest of us, and he has a dad.

"My dad is a retired small business owner who's always been quite conservative in his politics," he told me. "I think he leaned towards climate skepticism because that's what he would hear from the politicians he agrees with." As a result, conversations with his father have been difficult. In fact, the zombie arguments his dad kept bringing up, all the ones I talked about in Chapter 4, were what motivated John to create Skeptical Science. It's the educational website that enumerates and thoroughly debunks common climate change myths that I sent to my uncle back in Chapter 1. Did this massive avalanche of scientific data convince John's dad? By now, you can probably guess: no, it did not.

John's father lives in Gympie, in rural Queensland, Australia. In 2009, the government wanted to encourage people to install solar panels on their roofs by offering incentives. You would get paid twice what you were paying for electricity for any power that you sent back to the grid. "I had solar installed at my house," John said, "and when I mentioned this program to my dad, he was initially resistant, probably because I was the one saying it and it seemed like a green-y thing to do."

Then one day he came to John and said, "I've crunched the numbers and it's a financial no-brainer to get solar power." His dad is a thrifty man, a fiscal conservative. Saving money is one of his core values, so he had sixteen panels installed on his roof, a three-kilowatt system. Every time he received a check from the electricity company, he would call John to tell him about it. "He estimated that every year the panels saved him twelve hundred dollars. He never paid another electricity bill in that house," John said. Having solar panels wasn't just consistent with his values—they were turning him into an even better version of himself, thriftier and even more conservative (in the true sense of the word). They enhanced rather than challenged his identity.

A few years later over dinner, John's dad told him, unprompted, in the course of their conversation, "Of course humans are causing global warming." John nearly fell out of his chair. It was the last thing he'd expected, having gotten nowhere on this subject with his father for years. When John asked, dumbfounded, "What changed your mind?" he was even more surprised by his dad's response: "What are you talking about? I've always thought this."

As a cognitive psychologist, John felt like he was living in a surreal version of one of his own studies. His dad had denied the science for years, and now he was denying that he'd ever denied it.

Given the importance of solution aversion in driving denial, John suspects that when his dad shifted his behavior to be more climate friendly—even though he did it for financial reasons—it precipitated a change in attitude. His dad's perception of who he was had been altered, and at such a fundamental level that he literally couldn't recall that he'd changed.

Our actions reinforce, deepen, and can even irrevocably alter our sense of who we are. Not only that, but what we do changes others, too. And the contagion of seeing others act, says behavioral economist Robert Frank, can spread "more like outbreaks of measles or chicken pox than a process of rational choice." The difference is that, unlike coronavirus, behavioral contagion can be a good thing.

TRACKING SOLAR CONTAGION

In 2015, two geographers noticed solar panels popping up on houses in their small U.S. state of Connecticut. Curious, they set out to see if they could figure out what predicted who had them. Would they be on richer homes? Or in areas with higher population density?

Early adopters of solar panels tend to be people who are interested in innovative technology, who find an installer they trust, and who think having solar panels will benefit them. But once an early adopter made their choice, the geographers found, a cluster would spring up around them. Having solar panels on a house near you, where you could see them and talk to a real live person who had them, it turned out, was the biggest predictor of whether you'd get them yourself. Why? Because it brought down the "cost" of information. You didn't have to go somewhere or find a new person to talk to; they were right there beside you and, like John's dad, ready and eager to bend your ear with all the information you needed to make the same choice they had. Soon the Connecticut study was being replicated—in Sweden, in China, and in Germany, where they actually

put a number on it: rooftop solar installations were most influential, they
found, on neighbors who lived within one kilometer. A Swiss study went
one step further and recommended that such hot spots be deliberately
created to spur adoption.

By 2020, 21 percent of homes in Australia had rooftop solar. Government mandates in the state of California and the city of South Miami
mean that most new homes there must have rooftop solar, too. In 2021,
the global rooftop solar industry was valued at nearly $40 billion world-
wide and expected to pass $80 billion by 2027.

From where I live in West Texas, all the way across to southern California, we receive the most solar energy of anywhere in the U.S. Rainy
days are as rare here as a sunny winter day in Vancouver or London. We
get so much sun that, using presently available technology, I've calculated
you'd need to cover little more than a square area one hundred miles
per side with solar photo-voltaics—which would fit easily right between
Lubbock and Amarillo—to supply the entire U.S. with electricity. That's
similar to the area currently being used nationwide for maple syrup production, golf courses, and airports combined.*

Five years ago, Texas wasn't even in the top ten of commercial solar
energy producing states. Fast-forward to 2020 and every few weeks, it
seemed, there was a new update on solar in Texas. In November, a $1.6
billion Invenergy project near Dallas was announced to supply A&T and
Google. With seven thousand megawatts of installed solar (and an estimated fifteen thousand more on the way in the next five years), Texas is
now the second largest solar producing state in the U.S.

Similarly, five years ago I didn't know of more than one or two homes
in our city of over a quarter million people that had rooftop panels. In
Texas we didn't have enough early adopters, it turned out, for clusters
to form. Then, just before Christmas two years ago, I got a notice saying that my credit record had been pulled. (This may seem like a non

---

*Of course you wouldn't want to generate a whole country's electricity in one location;
this example is just to illustrate the potential of Texas solar and the fact that it's really not
a lot of land we're talking about.

sequitur but stay with me here.) I knew I hadn't applied for anything, so it could only mean one thing: I'd had my identity stolen. I frantically called my husband—since we share our credit cards—but before I could freeze them, he said, "Don't. It's okay."

"What do you mean, *okay*?" I asked.

"I can't tell you; it would ruin the surprise."

"*What* surprise?"

It took a few days before he relented and told me: being a tech-savvy person who likes saving money, which fits the profile of an early adopter, he'd crunched the numbers, completed the negotiations, and bought us solar panels for Christmas. He had the added advantage of knowing how happy I'd be, and I was. I literally teared up with joy when he told me of the surprise gift he'd planned, even when he confessed that the electrician had accidentally put his leg through our bedroom ceiling while installing the panels. And here's the best part: he'd purchased them from a local company that uses panels manufactured by a company called Mission Solar in San Antonio. The last time oil prices tanked and many of the people working the rigs in West Texas lost their jobs, this company took in oil workers and retrained them to manufacture solar panels. Mission Solar is part of the just transition movement—and because we bought from them, we are, too.

Having the panels, and knowing who we got them from, gives me something to talk about that I love. Now three other people we know have them, and within six months, there was another set of solar panels on an unknown neighbor's house, one block over from us. Today, I know a local business and a church that are considering them, too.

Yes, solar panels cut my carbon emissions, but they also make me feel empowered, as if what I do matters. They gave me a sense of *efficacy*.

BUILDING EFFICACY

Stanford psychologist Albert Bandura has been studying human behavior since before I was born. In 1977, he proposed—and proved—that people change their behavior if they feel *self efficacy*, which he defined as "the belief in one's capabilities to organize and execute a course of action." "Feel" is not really the right word, as efficacy is not technically an emotion. Rather, psychologists refer to it as a cognitive process. Perhaps it's more accurate to say that if you *think* you can do something, like hiring your neighbor's installer to put some panels on your roof, you're more likely to. And if you think what you do will make a difference (for example, you'll save money and feel good about yourself), that's even better.

Surveys of people in different countries show that people's sense of efficacy when it comes to climate change is not high. Even those of us who are concerned about climate change often feel as though we aren't able to make much of a difference. In the U.S., one survey showed that over 50 percent of Americans feel helpless when they think about climate change. Another survey found that more than 50 percent "don't know where to start" when it came to climate action.

Some barriers arise because we simply don't know enough. We go to the home improvement store and stand in the lightbulb aisle, staring at unfamiliar-looking bulbs. What type of LED should I use to replace the old 60-watt incandescent in my favorite reading lamp? The last one I tried made me feel like I was in an interrogation room.

In other cases, we might not even be able to access the information we need to make the best decision. Eating locally grown food typically has a lower carbon footprint, so you might download an app that tracks "food miles"—how far your food has to travel from farm to fork. But there are exceptions: food produced nearby but transported by truck may have a larger footprint than food produced far away but transported by rail. How do you sort out which is which when you're at the grocery store?

And then you find out that *what* you eat is far more important than how it gets to you, so you cut out meat from your diet. But then you learn that other animal products, like yogurt and eggs, are nearly as bad. You cut those out, too—and then you realize that your dog is eating even more meat than you ever did. In fact, the food consumed by your average-sized dog, not a particularly large dog, just an average one, has an average annual carbon footprint almost a quarter that of a typical passenger car's. What!?

Other barriers arise because we have different priorities. We may live in a big city with access to public transit. We still prefer to drive to work because it makes it easier to drop our child off at school, or maybe we just like to spend our time alone. We've read articles about people who can fit their annual nonrecyclable waste into a single jar (a *jar*?), but the time required to eliminate all that waste from our lives feels prohibitive and overwhelming.

Still other barriers are logistical or financial. We may dream of having an electric car, but we simply can't afford it. Or we'd like to stop flying—or fly only in biofuel-powered planes—but our job requires at least some in-person travel or we live across the continent, or even across the world, from our families.

But by a mile, the biggest barriers are emotional and ideological. We may be concerned, worried, or alarmed about climate change, but we don't have a sense of efficacy. As clinical psychologist Rubin Khoddam points out, we humans constantly fall victim to the motivation trap, waiting until we feel like it until we act. In fact, he says, "valued action," meaning action that is consistent with your values, "comes first," and motivation follows.

Then in terms of ideology, when solutions are presented as pertaining to the more liberal end of the political spectrum, conservatives see them as oppositional. Not only can they not do them, they don't want to. But what if action turns out to be not only doable, but consistent with their values, just like John's dad? All of a sudden their objection to the issue itself evaporates, because they can be part of the solution now, rather than being part of the problem.

WHY ACTION EMPOWERS

What builds our sense of efficacy when it comes to climate action? Research is still emerging, but the bottom line is pretty intuitive. When you hear or see or learn about what the real solutions look like, and how many of them are already being implemented or will be in the near future, that can increase your efficacy. And when you see someone else do something or find out about something you can do in your personal life—or in the case of John's dad, something he'd already done—that increases your efficacy, too.

It's a true positive feedback cycle. When we feel empowered to act, individually and communally, that makes us not only more *likely* to act, but to support others who do. It's a very human response that has been identified again and again around the world. It also inoculates us against despair: young people who are anxious about climate change, one survey found, aren't paralyzed by it if they are able to act. People along the U.S. Gulf Coast who were affected by the *Deepwater Horizon* oil spill turned out to be less depressed if they were participating in the cleanup themselves and actively doing something about it. And in general, the more we do something, the more it matters to us and the more we care.

It's not about being a lone ranger, either. *Collective efficacy* is even more important—the idea that together, as a community, we can make a difference. That's why it's so important to seek out like-minded groups: other athletes, parents, fellow birders or Rotarians, or people who share our faith. Together, our actions add up; it's not just us alone anymore.

That's the premise behind Lisa Altieri's BrightAction, an online community platform in the U.S. Households can set goals and take action as part of their community: their neighborhood, their scout troop, their church, or their place of work. They can share information online or in real life, and even compete against other "teams." Who's in? Dozens of cities, from Palo Alto, California, to Albany, New York; all the Hawaiian islands; even Arizona State University and the Episcopal Church of America have hired Lisa to create customized communities where people can cut their carbon emissions together. And fostering a sense of collective efficacy by getting things done is what Citizens' Climate Lobby, or CCL, is all about, too.

Marshall Saunders was a former Shell Oil employee and real estate broker from Waco, Texas. In his own words, he "awoke to the climate crisis when he saw the [Al Gore] documentary *An Inconvenient Truth* [and] discovered the power that a well-organized, properly trained group of citizens could wield to make the world a better place." He founded CCL in 2007. By the time he passed away in 2019, it had over six hundred active chapters spread across sixty countries and every continent of the world except Antarctica.

In the U.S., CCL has created bipartisan climate solutions caucuses in the Senate and in Congress. In Canada, they successfully lobbied for a fee-and-dividend approach to the national carbon tax. In Slovakia, I've shared the stage with a CCL advocate who explained the benefits of this approach for a coal-dependent country. And as I've traveled, I've met business people, faith leaders, retirees, and even many academics—from literature professors to astrophysicists—who wanted to advocate for climate solutions but felt helpless until they found a local CCL chapter. There, together with other concerned community members, they could be encouraged and encourage others. People could grow their collective efficacy through learning more about climate solutions and having conversations and meetings with their neighbors and their elected officials to advocate for change.

TALKING CLIMATE SOLUTIONS WITH POLITICIANS

David is a physician. He lives in Utah and grew up in the Church of Jesus Christ of Latter-Day Saints, so he understands the faith that motivates so many in that state. He also knows what the thick blanket of smog hanging over Salt Lake City looks like and how it affects the health of the kids and adults who live there.

When David retired, he knew he wanted to continue the fight for his patients' health. He also knew that fixing air pollution and climate change was key to that—so he joined CCL. His local chapter asked if he'd be willing to reach out to his Republican state senator. "Sure!" he replied, with no idea of what he might be taking on. But my favorite thing about CCL is how they teach people to approach conversations, even with potentially hostile politicians: with gratitude for what they are doing for their constituents, and the attitude that we are all on the same page and trying to do the right thing.

The first time David met with his state senator, the senator was cordial but cautious. He didn't want to talk about climate change, but they found common ground on air quality because the senator was an avid biker. He often rode his bike fifteen miles to the state capitol for his legislative work. So his next visit, David started the conversation with appreciation for the senator's desire to clean up Utah's air. Soon, the senator invited him to spend the day with him at the Capitol during the legislative session. By the second year, his relationship with the senator was so close that when David mentioned that a Christian climate scientist (me) was coming to town and would he like to have breakfast, the answer was an immediate *yes*.

Thanks to David, our conversation with the senator focused entirely on real problems and viable solutions. There was no discussion of thermometers or hoaxes—just genuine concern for those being impacted by climate change and air pollution and the challenges of accelerating the clean energy transition for the good of all. In some small towns in Utah, the entire community is built around the coal mine. If it's shut down, it will throw people out of work and devastate the town. So how to attract a new industry to town, and how to ensure it would offer training and jobs, was the question at hand. Cutting coal use doesn't just help with climate change, which is already hurting Utah's lucrative winter recreation industry: it also helps with the air pollution as well. The senator cared, and David cared, because they both cared about the place where they lived and the people they shared it with. It just made sense.

Marshall Saunders once said, "I used to think that the important people were taking care of the important problems. I don't think that anymore." He didn't necessarily mean that it's a bad thing. His epiphany was simply that leaders don't magically fix problems, even when they're important. What Marshall realized was that ordinary people share the power to fix important things—and indeed, are the best hope of getting things done.

No matter what our place in society, important problems don't get fixed until enough ordinary people mobilize to take action. It isn't only about what we accomplish ourselves: connecting with others imbues us with a stronger sense of collective efficacy and builds a network of like-minded people. Sharing our opinions and actions alters social norms, the informal rules that govern our behavior. This in turn makes us more likely to support politicians who want climate action and policies to reduce carbon emissions, more likely to speak out about the need for climate solutions, and more likely to be in favor of the changes required to address climate change at scale. It's like knocking over the first domino: action eventually changes us *all*.

WHAT I DO

"Eating organic is nice, but if your goal is to save the climate your vote is much more important."

DAVID WALLACE-WELLS, *THE UNINHABITABLE EARTH*

"We all try and solve a different part of the puzzle."

NATASJA VAN GESTEL, SOIL ECOLOGIST AND KATHARINE'S COLLEAGUE

Every year, I add two new low-carbon habits to my life. I don't do it because I believe my personal carbon emission reductions will make a difference. Even if all of us who care do our best, as I calculated in Chapter 13, our individual choices will never cut global carbon emissions to anywhere near the goal of the Paris Agreement. So why do I adopt these new habits?

First, because it's the right thing to do. Even if its impact on the world is meaningless, it's important to me to feel like I am doing my part. I also do it because research has shown that climate scientists who take their own carbon footprint more seriously are perceived to be more credible messengers and are more effective advocates for others taking action and supporting climate-friendly policies. That's no surprise; no one likes a hypocrite. But most of all, I do it because it inspires me, and it helps me inspire others, too. It knocks over the first domino, reminding me that action is possible. It gives me something to talk about, and it builds efficacy in others as I share with them (without lecturing or hectoring) what I've been doing and how I feel about it.

STEPPING ON THE CARBON SCALES

When you're going to lose weight, the first thing you do is step on the scales to see what you weigh now; then you set a goal of where you want to be; and lastly, you determine what you're going to do to get there. Count calories? Hire a personal trainer? Put a lock on the fridge?

In the same way, the first step to cutting your carbon is to step on the carbon scales. And when I did this some years ago, it showed me that my travel was the first thing I had to tackle. The result was the virtual-and-bundle travel policy I described earlier. Now, I've put it on my website so everyone who goes there can see it. Pre-COVID, I often heard from people who'd never tried a virtual talk before but because that was the only kind I would do, they were willing to give it a try. Now, of course, nearly everyone is on board.

When it's time to make big purchases, I also factor in my carbon. When I replaced my old hybrid with a plug-in electric car, we had to charge it outside the house until we got an outlet installed in the garage. We lived on a small cul-de-sac where all the neighbors would wave from behind the closed windows of their SUVs as they passed. But when they saw the plug-in, they were astonished. Every neighbor would stop, get out of the car, and ask somewhat incredulously, "What is *that*?" On being told it was an electric car, they'd ask, "Where did you get it?" and "Does it have a gas pedal?" and "How much does it cost to charge?" (a lot less than a tank of gas) and "Can I take a look?" Next time, they'd roll down the window and lean out. "I *love* your car," they'd say with a big smile. It was clearly the first one they'd ever seen: and they wouldn't forget it.

Even small changes can add up: washing your clothes in cold water instead of hot (which uses five times less energy and keeps your clothes from fading), or using smart power strips with your home computer and entertainment systems to reduce "vampire" load from devices in standby mode (some $19 billion, or $165 per household, of energy is wasted on this in the U.S. each year). One year, I finally sat down and figured out how many Kelvins I wanted my light bulbs to be so I could replace all our incandescent bulbs with soft white and warm LEDs. They last for years rather than months and use a fraction of the energy to run.

YOUR FOOTPRINT IS WHAT YOU EAT

As much as a third of the food grown and raised on the planet goes to waste. At the global scale, it's estimated that 8 percent of human emissions of heat-trapping gases, primarily methane, are the result of that waste. If food waste were its own country, it would be the third biggest emitter annually today, after China and the U.S.

In developing countries, it's food that doesn't make it to market or isn't preserved before it decays. In wealthy countries, a lot of food is wasted before it ever hits your shopping cart. It's rejected at the farm gate as too misshapen for supermarkets, or plowed into the ground because of a decrease in demand. A lot is wasted after we get it, too. In the U.S., people throw away enough food from their own plates or refrigerators each day to fill a ninety-thousand-seat football stadium.

The food rescue organization Second Harvest estimates 58 percent of the food produced in Canada is either lost or wasted. About half of that occurs during processing, the other half during consuming—and half of it is avoidable. Food is valuable so an organization called Flashfood partnered with Loblaws, a popular Canadian grocery chain, to create an app that sells food approaching its "best before" date at a discount of 50 percent or more. According to Flashfood's founder, Josh Domingues, as of 2020 the app has saved over nine thousand tons of food from the landfill—and he's a customer himself. "It'll actually dictate what I buy for dinner," Domingues said.

I don't live in Canada anymore, but there's still a lot I can do. Instead of one big grocery haul every two weeks, I make two or three small trips a week on the way home from campus. It's quicker, easier, and it's not hard to figure out what to cook every night; there are only two choices and the veggies are always fresh. I don't need our extra freezer anymore, so I sold it and filled the empty space with racks where I now hang laundry instead of throwing it in the dryer.

I also started looking at what we ate. Eating lower down in the food chain produces fewer heat-trapping gases, especially methane. On Earth today we are raising more than 30 billion land animals to consume. Livestock emissions—from deforestation, feed, fertilizer, and yes, cow burps and farts—account for 14 percent of total global heat-trapping gas emissions each year. Per kilogram, beef produces the most: one hundred kilograms of greenhouse gases for each kilogram of beef produced. Chicken is just a tenth that, at ten kilograms CO_2eq per kilo of chicken. Eggs are five, and most fruits and vegetables are near or below one kilogram.

In fact, if animal agriculture were its own country, it would tie at number three for annual heat-trapping gas emissions. That's why, in countries where people eat a lot of industrially produced meat, adopting a plant-rich diet diet is one of the most impactful steps we can take as individuals to reduce our personal emissions. Eating less meat means fewer animals belching methane, and it also saves us money and improves our health. When our family does eat beef, we buy from a local farm. Free-range grazing is more humane, sequesters carbon in the soil, and promotes animal health. There are a lot of invasive species here in Texas, too, from wild hogs to deer, that have to be regularly culled. For over thirty years, Broken Arrow Ranch has been working with wildlife ecologists and ranch owners to safely butcher and process the meat, turning it into steaks, sausage, and ground meat that would otherwise go to waste. And as a wider array of substitutes for milk, cheese, yogurt, butter, and even meat become available in most supermarkets in rich countries, making these changes increasingly doesn't mean going without. My son became such a fan of Beyond Meat that when his grandparents were taking him out for burgers, he insisted they go somewhere that offered it as he preferred it to beef. In Toronto, it wasn't too difficult to find. My dad tried one, and now he's on board, too.

Feeding pets is more of a problem, especially as there is a trend toward feeding dogs and cats fancy "human grade" meals. Instead, look for pet foods that take a "snout to tail" approach, serving parts of animals that humans don't typically consume. A Futurra start-up called Lovebug and another by Purina called RootLab are developing and selling cat and dog food made from innovative proteins with a small carbon footprint, such as invasive Asian carp and cricket meal. While you might not be particularly excited to tuck into a cricket burger, Fido and Fluffy—or Timbit and Dr. Evil, as we call them—aren't likely to recognize the difference.

EVEN THE BEST LAID PLANS CAN FAIL

There's something I can't do much of right now during these COVID times, though, and that's recycle. Lubbock's municipal recycling program handles a very limited range of items, and much of it just ends up in the dump. So for years, I used our university housing department's home-grown program. It is the work of one woman, Melanie Tatum, who carefully sourced and pieced together a host of different recycling options, from glass and plastics to electronics and Styrofoam. Even better, she used the proceeds from the recycling to support student workers, so my garbage was helping students.

But then coronavirus hit, and our campus shut down. For weeks I faithfully piled up our recycling in our garage, hoping against hope that the campus would open back up again soon; but no such luck. I even found myself considering how much time it would take to load up my car and drive all my recycling to the nearest city with a decent recycling program, six hours away. I recognized that would likely consume more energy than just tossing it, but I still had to hide my eyes while my husband took the bags to the dumpster. The week afterward, a student wanted to talk to me: she had the same dilemma and didn't know what to do. This same scenario was replicated across the world. As household garbage increased under coronavirus—by as much as 25 percent in the U.S.—supply chains for recycling were disrupted. Some cities had no choice: the recycling went to the dump.

This illustrates how we may try our hardest to live up to our ideals, but sometimes the more we try, the more hopeless it starts to feel that we can ever succeed. You might even run into the work of my friend Kim Nicholas, a sustainability scientist at Lund University in Sweden. She's calculated that the highest-impact personal carbon emission reduction anyone can make is not to have a child (and the second is to go car-free). If you're a parent like me, that ship has already sailed. So should I just give up?

KEEPING YOUR EYE ON THE BALL

This guilt-based system of believing our individual choices are what's needed to save the world will exhaust us. And when we're exhausted, when we feel like we've done everything we can and it still wasn't enough, it's more tempting than not to just throw in the towel and, to paraphrase the prophet Isaiah, think "I might as well just eat, drink, take great vacations, and drive a giant SUV, right? If we're all going down, why not enjoy the trip?"

So if you feel that, remind yourself—as I have to remind myself, too—that what really counts, what really carries the weight, is when we know we can act, and we share that sense of efficacy with others. That's how social contagion begins.

"Look at the world around you," Malcolm Gladwell says in *The Tipping Point*. "It may seem like an immovable, implacable place. It is not. With the slightest push—in just the right place—it can be tipped." And in what direction do we want to tip it? As my colleague Michael Mann wrote in his 2019 *TIME* essay, "Lifestyle Changes Aren't Enough to Save the Planet":

> We don't need to ban cars; we need to electrify them (and we need that electricity to come from clean energy). We don't need to ban burgers; we need climate-friendly beef. To spur these changes, we need to put a price on carbon, to incentivize polluters to invest in these solutions. . . .
>
> Focusing on individual choices around air travel and beef consumption heightens the risk of losing sight of the gorilla in the room: civilization's reliance on fossil fuels for energy and transport overall, which

accounts for roughly two-thirds of global carbon emissions. We need systemic changes that will reduce everyone's carbon footprint, whether or not they care.

That's why the single most important thing that I do, and that you can do, too, has nothing to do with solar panels, or food, or recycling, or lightbulbs. The most important thing every single one of us can do about climate change is talk about it—why it matters, and how we can fix it—and use our voices to advocate for change within our spheres of influence. As a parent, child, family member, or friend; student, employee, or boss; shareholder, stakeholder, member, or citizen: connecting with one another is how we change ourselves, how we change others, and ultimately, how we change the world. It's contagious.

20

WHY TALKING MATTERS

"If norms lead people to silence themselves, status quo can persist. But one day, someone challenges the norm. After that small challenge, others may begin to see what they think. Once that happens, a drip can become a flood."

CASS SUNSTEIN, *HOW CHANGE HAPPENS*

"The majority of people were eager to commence our discussion."

HOWARD FROM BRITISH COLUMBIA

Climate solutions are complex and multifaceted. Our response to the challenges climate change poses to our world, our identity, and our way of life are even more so. It's taken a whole book even to begin to untangle them. But the first, crucial step forward is simple. For you, for me, for every single person reading or listening to this book, there is one simple thing that we can all do:

Talk about it.

A year or so ago I was reminded of how powerful this can be. I'd just finished a talk at the London School of Economics and was heading up the aisle of the underground lecture hall when an older man named Glyn approached me. He said that he lived in Wandsworth, a borough of London, and had taken the train in specifically to hear me speak. He'd watched my TED Talk called *The Most Important Thing You Can Do to Fight Climate Change Is Talk About It*, and it had inspired him to have conversations about climate change with people in the borough where he lived.

I was amazed. Hearing that something I've done has made a difference—even just to one person—is why I do what I do. I sometimes get discouraged, and his words meant more to me than he knew. But Glyn wasn't done yet.

He'd started keeping a record of all the people who'd joined in with these conversations, he said. "Would you like to see the list?" he asked.

"Of course!" I said, surprised. I'd never heard anything like that before.

He reached in his leather satchel and pulled out a stack of papers. I'd been expecting about seventy or eighty names. But his list recorded over ten thousand names. Now it's upwards of twelve thousand (I checked back in with him before writing this). *Twelve thousand* conversations about climate change in a single English city borough, all because of one man watching one TED Talk about how important it is that we talk about why climate change matters to us and what we can do about it.

And that wasn't all. His borough had just voted to declare a climate emergency, he said—because of the conversations they'd had. Now, two years on, they've also divested from fossil fuels, invested in renewables, and just before COVID they announced they'd be spending £20 million on their new environment and sustainability strategy.

WHAT HAPPENS WHEN WE DON'T TALK

You can do what Glyn did: use your voice to talk about why climate change matters to you, here and now. Use it to share what you are doing, what others are doing, what they can do. Use it to advocate for change at every level—in your family, your school, your organization, your place of work or worship, your city or your town, your state or your province. Use it to vote and to inform decisions your school, your business, your city, and your country can make. Talk about it in every community that you are part of and whose values and interests you share.

Talking may sound simple, almost too simple. But here's the thing: most of us are not doing it. Even people who are alarmed and concerned about climate change tend to "self-silence" on the topic, says Nathan Geiger, a communications researcher. They want to speak up, and they know it's important, but they can't get the words out of their mouths.

Nathan decided to study environmental educators. These are people who are trained in communication and whose job it is to talk to the public. He found even *they* often hesitate to talk about climate change. And not doing so has repercussions for them; serious ones, he discovered. Many of them say they suffer from "severe psychological distress," he writes, "as a result of not being able to connect with others by discussing a topic about which they report concern."

How do the rest of us compare? According to polling data from the Yale Climate Communication Program, when people across the U.S. are asked, "Do you discuss global warming at least occasionally?" the answer was mostly *no*. Only 35 percent of people discuss it even once in a while.

What *do* we talk about? Things we care about. Our speech is the television screen of our mind, so to speak. It displays what we're thinking about to others, which in turn connects us to their minds and thoughts. So if we don't talk about climate change, why would anyone around us know that we care—or begin to care themselves if they don't already? And if they don't care, why would they act?

Don't be afraid of sounding like a broken record. We learn things from hearing them, again and again. As health and communication researcher Ed Maibach has been saying to anyone who will listen for the last twenty years, "the most effective communication strategies are based on simple messages, repeated often, by many trusted messengers." In other words, the eighth time you've said something, people will just be paying attention. What do people pay attention to most? In general, we tend to favor personal stories and experiences over reams of data or facts. In fact, when you hear a story, neuroscientists have found, your brain waves start to synchronize with those of the storyteller. Your emotions follow. And that's how change happens.

QUOTE SURPRISING SOURCES

Sharing a story about what an influential or a surprising person or group of people are doing about climate change can be a good conversation starter, and there are plenty to choose from. They can be religious leaders, such as the Pope, or the Archbishop of Canterbury, or the Dalai Lama, or the more than eighty Muslim leaders who signed the Islamic Declaration on Climate Change, or one of the more than twenty thousand Young Evangelicals for Climate Action across the U.S. They can also be politicians whom people trust—although these seem to be an endangered species of late. And they can even be celebrities. Although you may be skeptical of someone who lives a high-flying life and then touts the need for climate action and carbon cuts, at least they are using their large platform to raise awareness for good.

Business leaders are often perceived to be hardheaded and practical, with an eye to the bottom line. So when someone like Alan Jope, CEO of the large multinational corporation Unilever, says that "any company* that hasn't already got a net-zero ambition of some sort should be ashamed of themselves," as he did on a podcast he and I did with the World Economic Forum in September 2020; or Bill Gates, founder of Microsoft, sets a net zero goal for Microsoft, commits his fortune to moving the world beyond fossil fuels, and says, "As awful as the COVID pandemic is, climate change could be worse," many who respect their business acumen and financial success will listen.

Military officials can speak with authority to threats, and hearing what they think can often carry weight. In 2013, for example, when North Korea annulled their 1953 armistice with South Korea, the chief of U.S. Pacific Command, Admiral Samuel Locklear III, was asked what he thought was the biggest threat to the security of the Pacific. He didn't say North Korea or a nuclear attack. Instead, he said, "Climate change." Why does the military think climate change is a threat? "Drought and severe storms are triggering mass refugee migrations while devastated

---

*Note he's talking about shaming companies here, not shaming individual people.

areas could become breeding grounds for terrorists," says retired air force general Ron Keys. "We need to protect ourselves from these risks. This has to be everybody's fight."

Doctors and health-care professionals are widely trusted on health-related issues, and that's exactly what climate change is. Amanda Millstein is a primary care pediatrician in California. She treats kids whose asthma flares up when air pollution spikes during heat waves and wildfire smoke turns the skies an apocalyptic orange. "Climate change is about health, and specifically the health of our children. *Your* children," she says. So, speak up and advocate for change, in your community, your school, your company, and beyond. She continues, "COVID will eventually end. There is no vaccine for climate change."

SHARE WHAT SCIENTISTS SAY

As much as we complain about the weather, it *is* one of our most frequent topics of conversation, and we usually trust our local forecaster. It turns out their opinions on climate change can carry some weight, too. But first, some of them have to be convinced. In a survey he ran in 2010, Ed Maibach found that only a third of TV weather forecasters in the U.S. accepted that climate change is human-caused. So he teamed up with the American Meteorological Society and Climate Central, a nonprofit climate news organization, to change that. That same year, they founded Climate Matters. It's a training and resource program that helps meteorologists learn about and report on local climate impacts. They began with one man: Jim Gandy, a broadcast meteorologist in Columbia, South Carolina. Ten years later, nearly one thousand media meteorologists participate in Climate Matters across almost five hundred local television stations.

Did it make a difference? Yes. In media markets where the weather forecasters were part of Climate Matters, people's awareness of the risks posed by climate change increased, even after just listening to a short six-minute segment. This program also changed the forecasters' minds. A follow-up study ten years later found that the majority of

U.S. broadcast meteorologists now agree that climate is changing and humans are responsible.

You can also tell people what scientists are saying. After all, we are the ones studying this thing, and we know that climate change is real, humans are responsible, the impacts are serious, and the time to act is now. Many scientists who engage the public have already been convinced of the severity of climate change for some time, and are speaking out. There's Canadian geneticist David Suzuki; Australian paleontologist Tim Flannery; American astrophysicist Neil deGrasse Tyson; and British primatologist Jane Goodall, who inspired me as a child. I curate a list of over three thousand scientists on Twitter who study climate, and I get questions from other scientists all the time about wanting to do more outreach.

Scientists care about climate change, and most of us want to talk about it. But even with all that we know and all the passion we have about this topic, we scientists aren't the most effective messengers on climate change. We're number two.

YOU'RE THE BEST MESSENGER

As effective as religious leaders or physicians or scientists can be, it turns out that the very best person to talk about climate change, the most trusted messenger when it comes to contentious and divisive issues, is not them. It's not me. It's *you*.

Yes, you. Someone who understands why this issue matters, who shares the same values, who cares about them—*you* are the perfect person to have this conversation with the people in your life.

Immediately, the "buts" raise their hands. But I'm not a scientist. But I don't know enough about it. But it's too overwhelming. But I can't cope with another depressing, frustrating conversation. But I'm not the right type of person to talk about this. Someone else can do it.

All of these "buts" are based on one big misconception: that the only way to have a conversation about climate change is to explain—or argue about—the science and to overwhelm people with the avalanche of bad news, starting with the most Dismissive person you know. Trust me, I've tried that approach. If talking more science would fix this, I can talk science with the best of them. As for Dismissives, though they may be the loudest voices, hundreds of attempts have taught me that conversations with the seven-percenters are largely fruitless.

Does that mean that the facts and the science don't matter? Of course not. As I argued in Section 2, facts explain how our world works, and most of us like to know that. Facts also make us sit up and pay attention. They address the questions we have and provide solid answers to myths we might have heard.

But facts about the science are not enough to explain why climate change matters and why it's so urgent that we fix it. We need more. We need to understand how climate change matters to us, personally, and what we can do about it in our own lives. And you, not I, are the expert on that.

HERE'S HOW TO START

Often at this point I hear, "I'm willing to talk, but where can I find people to talk to? Everyone I know agrees with me so there's no point talking about it."

I strongly suspect that's not true: if we don't talk about climate change much, how do we really know how people around us feel about it? Maybe they are Concerned or Alarmed, but they are anxious and don't know what to do. Maybe they lack a sense of personal or communal efficacy. Maybe they are Cautious, and they're not sure why it matters to them. In any of those cases, talking is a good idea.

Behavioral scientist Meaghan Guckian has some ideas. She noticed people who cared about it weren't talking about climate change and decided to devote her doctoral dissertation to figuring out what might get them to open up.

Create opportunities to interact with people, she says. You'll never know what people really think about climate change unless you ask.

As I was originally writing this very page, someone who works at an environmental nonprofit organization in Canada pinged me. "Hey Katharine," she said, "your TED Talk inspired our friend Howard to talk to strangers about the climate crisis. Here's his story!"

Howard had recently retired from being the director of Youth Justice for British Columbia's Public Service Department, but he didn't want to stop helping people. So he decided he'd talk to people while on his weekly runs through the grounds of Royal Roads University in Victoria. It's a gorgeous campus built in an old growth forest of towering Douglas firs and red cedars.

Being nervous at approaching total strangers (as we all would be), he took a leaf out of the book *How to Have Impossible Conversations*. He posed a simple poll to them: on a scale of 1 to 10, with 1 being "Climate change is not an important issue in the world" and 10 being "Climate change represents the greatest challenge facing the Earth today," what number would you choose? Sometimes he'd also ask if there were anything that might make them change their mind.

Only a few people gave brief responses, he said. Most were eager to talk. Many gave a score of 8 or more, some even choosing 11 or 12. "When it came to families, children often prompted their parents to select a higher number," he said. Others dug deeper, disclosing their personal angst over their dependence on the fossil fuels that are causing the problem, the fact that they'd chosen not to have children due to the uncertain future, their conflict with family members employed in the Alberta oil patch, or how climate change was exacerbating other areas of personal stress in their lives.

This experience taught Howard many things—and if we all engaged in a similar exercise, we'd probably see them, too. First, he learned that people were eager to talk. Second, it was possible to disagree but remain respectful and constructive—easier, often, when you were total strangers rather than close family members. And lastly, everyone had something to share—concerns and solutions, too. "I was able to use what I learned to engage family members and friends in climate crisis conversations," he said, "which to date have gone well."

WHY THESE CONVERSATIONS MATTER

Look for opportunities to work together with people on climate solutions. This does double duty. It helps us understand that—both individually and collectively—we can make a difference. But while you're working with like-minded people, also talk with them. Discuss why climate change matters, what you're worried about, how what you're doing can make a difference, and how you might bring more people on board.

Talking about climate does matter; the results can be very powerful. In social-science-speak, your response efficacy is high. Connecting with people over genuinely shared values reaches directly into our hearts, past the barriers of "them" and "us" that we've erected. We can identify with one another over something that matters to us deeply and defines who we are. That makes it the perfect place to start the conversation. But that's not all.

Research by another climate communication specialist, Matthew Goldberg, has also shown that the simple act of having a conversation triggers a true positive feedback effect. The more we know about how climate change affects us, the more concerned we are. The more we worry, the more we talk about it. And the more we talk about something, the more conscious we are of the need to act—and aware of the mountains of things others are doing already and the millions of hands already rolling that boulder down the hill.

ANSWERING THE BIGGEST QUESTION

"Okay, I can see that this is important and I'm willing to give it a try; but where do I start? And what do I say?"

This is the *number one* question I get nearly every day, from almost anyone, anywhere. They'd like to talk about climate change. It's at the forefront of their mind. But, as Nathan's research shows, even if you feel that way, often you don't think you can. Your sense of personal efficacy may be low: you don't feel equipped to have a conversation that might take a deep dive into some science you're not familiar with. And your sense of response efficacy may be even lower. Previous conversations haven't been a pretty sight or yielded any positive results. Many have ended in frustration, or conflict, or maybe just depression, as participants agree it's a problem, but what can we do? Nothing.

I have good news. There is a way to talk about climate change that works. You don't need a PhD in climate science. You don't need a bulletproof vest. And you don't need any antidepressants, either. In fact, chances are you'll know more afterward than you did before; you'll have a better understanding of the person or people you're talking to than you did earlier; and you'll be encouraged rather than discouraged by your conversation. So what is this secret formula? It's this:

Bond, connect, and inspire.

BOND, CONNECT, AND INSPIRE

"Most people do not listen with the intent to understand. They listen with the intent to reply."

STEPHEN COVEY, *THE 7 HABITS OF HIGHLY EFFECTIVE PEOPLE*

"I never knew what to do about climate change before. But now that I know food waste is a big part of it, I'm going to make sure we eat all our Christmas leftovers!"

TORONTO CHURCHGOER, AFTER KATHARINE'S TALK

A few years ago I was attending the Christians in Science conference at Queens' College Cambridge, in the U.K. I'd just given my talk on climate science, impacts, and why it matters to Christians and the conference was on a tea break. I was sitting outside in the courtyard with a group of women who were asking me about the gendered abuse I receive as a climate scientist, when a younger man I hadn't met yet strode up to us. One glance at his face and I could tell: he was *ticked*.

An engineering professor from a university in the south of England, Tom was incensed by the concept of certainty, the idea that scientists could know the climate was changing. He also disagreed with my assertion that yes, scientists really had checked every other option and humans were responsible. Our conversation got off to a rocky start and deteriorated from there. It left me with the sincere hope I would never see him again—and he probably felt the same way. I still remember the horrified faces of the women around the table after he stormed off. Our conversation had included an unexpected real-life illustration.

The following July, I wasn't planning to attend any conferences. But I was in Toronto, and the Canadian version of the same group, called the Canadian Scientific & Christian Affiliation, was meeting at a nearby university. My dad was giving a talk, so he invited me to come along for the day. The session went well, but over lunch we were ambushed by a retired astrophysicist who had disagreed with what my dad had to say about creation care and climate change. Like so many of the retired engineers whose large manila envelopes arrive in my mailbox with regularity, he was devoting his golden years to showing why those upstart climate alarmists from the last hundred years were dead wrong.

At one point, when the astrophysicist was railing against "those IPCC scientists" and how they lied, I leaned across the table to him and said, "Hang on. Don't you realize I'm one of 'those climate scientists'? You are literally talking to one. Do you think I'm lying about this—and lying about being a Christian, too? Or do you truly think I'm a certified idiot—with a degree in astrophysics—who has absolutely no idea what she's talking about?"

From the flummoxed look on his face, it was pretty clear he'd never put a human face on "those climate scientists" before. When confronted with one in real life, he struggled with assigning to me the venality and dishonesty he'd labeled us all with. But it wasn't enough to disrupt his train of thought, and he soon headed off on a different tack.

As I talked about in depth in Section 1, there is no secret to a constructive conversation with a Dismissive. I don't think it's possible to have one, short of a genuine honest-to-God miracle. And although those do occur sometimes, in general the best you can hope for is to let them know you think they are wrong and disengage as soon as you can, unless you genuinely enjoy pointless argument. As fashion icon Coco Chanel famously said, "Don't spend time beating on a wall, hoping to transform it into a door."

This astrophysicist was clearly a wall, not a door. So, hoping to cool down on the walk back to the conference, I left my more patient dad behind to continue talking to him while I headed outside.

BE CAREFUL WHAT YOU ASK FOR

Distracted by the replay of the conversation in my head ("Was there any-thing I might have said that would have made a difference? No. Could I have responded more graciously? Definitely.") I walked out the wrong door. It slammed behind me; I tried to open it but it was locked—and there I was, on an unknown campus, with a dead phone and no idea where I was going.

I stood there for a little while, still steaming over the lunch conver-sation and trying to figure out which direction to head in. Suddenly, I heard the door behind me open again—salvation! Someone else was coming, maybe they'd know which way to go.

I turned around, and my jaw dropped, because there was Tom: the Tom from the tea break at Cambridge; the Tom whom I'd devoutly hoped I'd never see again; the Tom who, if I absolutely had to encounter him again, could it at least please God not have happened directly after an infuriating lunch with a Dismissive astrophysicist who'd just called me and all my colleagues venal liars to my face?

But there I was, and there he was, too, and there was not another human in sight as far as the eye could see. So I took a deep breath, and I could see him do the same. It was clear we were both mentally resolving to be civil and avoid the topic of climate change at all costs.

It was his first time on that campus, too, but he thought he knew the right direction, so we set off together: I in silence, still trying to swallow my lunch, metaphorically speaking, and he struggling to find a neutral topic of conversation.

Looking down, his eye alighted on my bag, where a pair of knitting needles were poking out. He brightened instantly and asked, "Do you knit? I knit, too."

I was surprised; while one sees the occasional woman scientist knitting at scientific meetings, this was the first time I'd had a man express interest. I replied that I was making a scarf for my mother's birthday in a few days.

"That's great," he said with genuine enthusiasm, "I do the same. In fact, I don't think we should buy presents for people we love. Who needs more plastic from China? We should make all our gifts. It's much more meaningful."

He went on to say that was what his family had done for Christmas last year, and when I genuinely told him how wonderful I thought that was, he warmed to his theme. He didn't just recycle, he said, he "upcycled"—building their furniture from packing pallets and refurbishing cast-off furniture he'd salvaged. His family lived in a small apartment near the city center, so they didn't own a car and hardly ever had to drive. He only traveled to one international conference a year that required a flight.

"And wasn't it my luck it was this one," I thought.

But listening to him, I realized that what he was describing was an incredibly thoughtful, sustainable, and low-carbon lifestyle. It was rich in what really mattered—family, friends, and life itself, not the goods and materials that we so often use to define ourselves. In fact, if all of us lived like he and his family did, we'd not only be healthier and in better shape, we'd be living true to our values. This revelation struck me with the force of a locomotive. Here was a fellow academic who wasn't on board with the science, but his life was an example to us all.

I absorbed this remarkable revelation and, as we reached our destination—because this unexpectedly harmonious conversation had carried us all the way across campus—I turned to him and told him sincerely, "I know we disagree about what climate science is telling us about our planet. But I'd rather everyone thought the same way you do, and lived the same way you do, than agreed with me but lived the way many of them do."

"Really?" he replied, clearly gobsmacked by what I had said.

"Yes," I said emphatically, "*I mean it.*"

He smiled, we walked in the door, and I never saw him again. But as I finished knitting my mom's scarf during the next session, I began to digest one of the most important lessons I've learned to date: that we don't really have to agree on the science, as long as we agree on something that matters *more*.

HOW TO BOND AND CONNECT

Whoever we are, we are human. And as humans, we have the power to connect with one another across many of the broad, deep lines scored across our societies and our psyches. We can't do this by bombarding people with more data, facts, and science showing they're wrong, or heaping on the judgment and guilt. Instead, we have to start with respect, and with something we both agree on: bonding over a value we truly share, and then making the connection between that value and a changing climate. By doing so, rather than trying to change who someone is, instead it can become clear that the person you are talking to is already the perfect person to care about and act on climate change. In fact chances are they probably already care, they just might not have realized why, or known what to do about it if they did.

The only reason I care about a changing climate myself is because it affects everything I already care about. My child. The future of our family. The places where we live—and how those places are being affected by more intense hurricanes, rising seas, stronger droughts, heavier rain. The food we eat, where we grow it, and how much it costs. The air we breathe and how clean—or dirty—it is. The economy, national security, justice, and equity, every single Sustainable Development Goal of the United Nations, the future of civilization as we know it. The list is endless. In fact, it's almost impossible not to find something that you can connect to climate change, once you start looking.

If you're wondering where to start bonding with someone and connecting on climate change, ask yourself, "Because of what we both care about, why might climate change matter to us?" A sense of place is always a key connection. If you both live along low-lying coastlines—in the eastern U.S. or the British Isles or Southeast Asia—you're already seeing flooding on sunny days. If you're farmers in Texas or East Africa or Syria, you've witnessed firsthand how climate change is shifting your seasons and amplifying your natural cycles of drought and flood, and hitting you right where it hurts, in the pocketbook. In southern Australia or western North America, bigger wildfires are putting your homes at risk. Do you live near the mountains? Shrinking snowpack is endangering your water supply. A northern country? All kinds of invasive species and pests are moving poleward as your winters warm. All through this book I've talked a lot about why climate impacts and climate solutions matter—to our health, to our hobbies and our homes, to the economy and our food and water, to people less fortunate than us, and more. Did anything there connect with you?

HOW TO INSPIRE

No matter how carefully you prepare, there will still be conversations that don't progress. But even some of them, as you saw with Tom, can take an unexpected turn when we start to talk about our lived experience rather than abstract data and facts. That's why I think the last step, to inspire each other with real-life, practical, and viable solutions, is the most critical.

Ask yourself: What solutions can I bring up that whoever I'm talking to might get excited about? Would they be interested in a free market solution to climate change from former congressman Bob Inglis's republicEn organization? Do they live in an agricultural area like Matt Russell where smart farming techniques to put carbon back in the soil might be of practical use to people? Would they like to hear more about Solar Sisters or Sulabh or other programs that are revolutionizing the lives of the energy-poor? Do they have a pet, and might they like to hear about cricket-based food?

Would solutions at the intersection of racial, gender, and indigenous justice catch their attention? Maybe they'd just like to hear about how you love your LED lightbulbs or your new plug-in car. Or perhaps there's a volunteer effort they could be involved with in the community, cleaning up a watershed or picking up garbage. Possibly they could come with you to an interesting presentation hosted by the local chapter of Citizens' Climate Lobby or the local university. You could start (or join) a creation care group at your church, or form a craft club to knit or crochet warming stripes.

The list of solutions is virtually endless. And being able to offer some up is essential for these conversations. The social science I talked about in Section 4 is clear: if we present people with a problem or a challenge, even one that has no politicization or controversy associated with it, but we don't offer an engaging solution, people feel disenfranchised and powerless. That is true whether we're talking about the simple fact that eating trans fats is bad for us, or discussing how we should be saving more for retirement, or speaking about the need to tackle climate change. If we don't offer a solution, things start to look insurmountable. Our brain's natural defense is to try its best to forget that the problem exists.

CONVERSATIONS I'VE HAD

Wherever I go, I have climate conversations. Each place is different, each uniquely vulnerable to climate impacts. But people in each place are the same: worried about the impacts they see today, anxious for the future, and full of ideas for how we can work together to fix this.

In California, I hear from colleagues who've had to evacuate their homes due to the latest wildfire. I meet graduate students who are passionate about protecting the urchins and whelks that populate the cold channel waters, and the local fishing and seafood industry whose jobs depend on them. I talk to schoolkids who have some of the best questions and ideas I've ever heard. (Is there any way we could hand the world over to them a little earlier?)

In Paris, I meet with Engie, a large multinational utility company that's planning to be the first carbon neutral energy provider in the world. I mention Project Drawdown, an inspiring organization that has researched and studied nearly a hundred different practical climate solutions. "Oh yes," says Jan Mertens, their chief science officer, thoughtfully. "I reviewed that list the other day and I think we're already implementing 40 percent of them." Because they are a pioneer rather than a follower, Engie is running into one challenge after another as they lead by example and commitment. But they are not giving up, and their work encourages me as well.

In Ireland, colleagues say, "I'm not sure if you've heard this before, but people here say, 'We're such a small part of the problem, nothing we do makes a difference.' How would you address this?" I laugh, because I hear this everywhere I go: Canadians and Norwegians, even Americans say, "What about China?"—while in China people say, "Per person we're emitting hardly anything compared to the Americans." We use the same arguments because as humans we think the same way. But the truth is, we can only fix it together. So I make sure to emphasize that in my presentations, with facts and figures specific to the country where I am.

In India, state hazard planners ask for information they can use to prepare for stronger heat waves, shifting rainfall patterns, and dwindling water supplies. They're worried about their cities, their farmers, their small villages, and the future of their states. I say yes, because I'm already working on the same type of data for Houston, Texas. There, my colleague Gavin recently called in to our meeting on climate impacts from a parking lot. He was stuck there for the day: he couldn't go home or to the office as the flooding—the fifth five-hundred-year flood event in five years—was too severe.

In my home city of Toronto, I give a sermon at one of the biggest churches in the area, connecting our Christian values to why we care about climate change. As people leave, I overhear one woman talking to another. "I never knew what to do about climate change before," she says, "so I did nothing. But now that I know that food waste is a big part of it, I know what to do—I'm going to make sure we eat all our Christmas leftovers!"

And at Stanford University, I meet with Kameron and a big group of other graduate students. They're all supersmart. They're mostly very worried. And they want to talk about climate change—but they don't know how. Kameron's already been through dozens of conversations with his dad, a conservative Christian who rejects the science, so he knows how frustrating it can be. But he also knows that the antidote to anxiety is action. So he's gotten the graduate students together to start building an app. It's called Climate Mind and it will help people figure out how to talk about climate change: what to bond over, what impacts to connect to what you care about, what science-sounding arguments people might bring up that you can briefly respond to, and finally what positive, constructive solutions you can talk about and engage in together. It's already online, if you want to check it out.

HOW TO BEGIN THE CONVERSATION

I've heard a lot of good talks, lectures, and even sermons in my life. They've taught me things I didn't know and given me ideas that I want to remember and apply to my life. While I listen, the ideas seem crystal clear. But when I go home and try to implement those changes in real life, I often draw a blank.

It feels like the ballroom dance class I signed up for with my friends in university. While the instructor was explaining the steps, the cha-cha seemed so straightforward. But five minutes later, I couldn't for the life of me figure out which foot went where . . . and someone was standing on one of them anyways.

I don't want this book to be like those sermons or dance class. So if you're feeling like this all made sense while you read it, but when you tried to put some of these ideas into action, your ideas slipped away like the greased watermelon we used to play water tag with at summer camp, this section is for you. Here's where to put your foot next.

You've probably already done step one. As you've been reading, you've thought of at least one or two people you could talk to who aren't Dismissive. It might be a colleague, a tennis partner, an old friend, someone in your congregation, a fellow parent at the PTA, or even a family member.* You've identified what you have in common: a value you both share, an activity you both enjoy, an aspect of your life you have in common. (If you need more ideas, go back to Chapters 2 and 3.)

The second step is to prepare. As the microbiologist Louis Pasteur,

---

*Be warned that family members can be the toughest conversations, because of the decades of baggage we tend to carry around! Like Howard in the park, maybe consider a few less personal conversations first.

who created some of the first vaccines, said, "chance favors the prepared mind." And by reading this book, you've already done a lot of preparation. You've read dozens of examples of how climate change is affecting things we all care about, here and now, and real-world solutions that people at every level, from kids to presidents, are doing every day. You know the good stuff and you're ready to share it.

Talking about climate change is an important thing to do with almost anyone you know, and I'd recommend starting with one or more people you feel comfortable with. But if you want to have even more of an impact, consider a conversation with a decision-maker: your local government representative, the principal of your child's school, your office manager, an administrator at your university, a leader at your church, the owner of your gym or yoga studio—anyone who can make concrete decisions regarding energy or other resource use on a larger scale than you can.

If you're gearing up to have a conversation with a decision-maker to advocate for a specific action or change, preparing takes on extra significance. You don't only want to identify something that they genuinely care about—their financial bottom line, their reputation, or taking care of their constituents—you also want to come armed with possible solutions. There's little point talking about a problem with a decision-maker if you don't have a proposal for how to fix it. They may not accept your solution, but at least it will start the ball rolling in the right direction and show that you're not just criticizing, you're interested in partnering with them on genuine, practical change.

Figure out what your school or your congregation might care about: maybe reducing the budget would be a good foundation for inspiring change. An energy audit could save money and reduce their carbon footprint at the same time. If you're asking your organization or university or city to commit to a climate goal, you probably want to know what others have done (comparisons with fierce rivals tend to work well) and whether there is a specific agreement they could sign or program they could join with others like them. If you're talking to your business about how you want to integrate sustainability into its supply chain, look for other companies that have already done this and succeeded. How did it impact their reputation, and their bottom line? If you're talking to a politician—well, you might not believe it, but they are human, too. Find out what are their top priorities and be prepared to show why the solution you're proposing fits right in. Consider who or what could make a practical change, and what would motivate them to do so. Their reason might be totally different than what motivates you, and that's okay. As Tom taught me, we don't have to agree on why, just what.

The third step is, don't be too attached to the outcome. Set reasonable expectations of what you can and can't achieve. You're not trying to convert anyone to a new religion, or even change their mind; that's not your responsibility. Your goal is to simply open the door, to start the conversation, to practice talking about what you care about and listen to what someone else cares about, too. In other words, you can plant a seed, you can fertilize and water it, but no matter how hard you try, you can't will it to grow. That's not within your power. As the Greek philosopher Epictetus said, "The chief task in life is simply this: to identify and separate matters so that I can say clearly to myself which are externals not under my control, and which have to do with the choices I actually control."

HOW TO HAVE THE CONVERSATION

Now you've stepped up to the edge of the diving board, it's time to jump in. Think about how you're going to hit the water or begin the conversation. A question is usually a safe bet. You might ask an open-ended question about what they think or feel. You could offer a scale, like Howard did with people in the park: "From 1 to 10, what do you think about X?" Or you could start with an interesting fact. As you now know, our brains are attracted to new information, so "Did you know X?" or "Have you heard about X?" is a good beginning, too. You could also, depending on how well you know the person, share how you feel: "I'm worried about X because Y." And you could talk about something good: "I'm excited about X" or "You wouldn't believe what I just heard."

Once you've hit the water, so to speak, the most important thing to do at that point is to listen. Listen, and then—as climate communicator Karin Kirk advocates—*keep* listening, because the longer you listen, the more you'll understand. As psychologist Tania Israel explains in her book, *Beyond Your Bubble*,

"For successful dialogue you need to try to understand people and help them feel safe and understood. . . . When people feel confronted or attacked, they shut down and become even more committed to polarized views. . . . Being respectful means not dismissing their views, values, or experiences."

Keep your ears pricked for things you can agree with and reinforce. And when you respond, do so empathetically. See if you can repeat back to them what they've said to you, emphasizing your points of agreement. Psychologist Renée Lertzman calls this process "attunement," literally tuning ourselves to our own and to each other's emotions, experiences, and perspectives. Stay tuned to how well you're connecting in the moment, too, she says, so you can course-correct as necessary. As Jonathan Haidt says, "You can't change people's minds by utterly refuting their arguments. . . . If you really want to change someone's mind in a moral or political matter, you'll need to see things from that person's angle as well as your own. *Empathy is an antidote to righteousness.*"

HOW TO END THE CONVERSATION

This leads right to the next step, which is: know when to stop. If your emotions are rising to where you can't engage respectfully anymore, or you sense yourself trying to push back or judge the other person, or they're doing the same to you, it's time to move on or, if necessary, gracefully retreat. Remember, you're just trying to open the door, not convince someone to renovate their house—and you're certainly not trying to renovate it for them.

You're not quite done yet. The final step is this: learn from your conversation. Reflect on what you heard. As Climate Outreach's helpful manual, *Talking Climate*, says, every climate change conversation you have is valuable. "See the experience as a way to learn about how others think about climate change, about the topic itself—and about how to have a good conversation. Every climate exchange is a small experiment!" Keep going, they say, and keep connected.

Despite our daily frustrations, we know that in this increasingly polarized, divided, and fractured world, there's still far more that connects us than divides us. So whoever you are, wherever you live, look for opportunities to have a conversation. Be confident you can make a difference: you can, even if you never see or hear the results yourself. You don't know what consequences your conversation could have, now or down the road. Prepare with information on how it matters and what real solutions look like. Decide to do it: make a commitment to have a conversation and carry through. Listen, empathize, and learn from the experience. And if something works well, far beyond what you imagined—or if it fails even more spectacularly than my conversation with Tom and you'd like to share that—tell me about it! I'd love to hear your story.

22

FINDING HOPE AND COURAGE

"It is a magnificent thing to be alive in a moment that matters so much."

KATHARINE WILKINSON, TED TALK

"Hope has two beautiful daughters; their names are Anger and Courage. Anger at the way things are, and Courage to see that they do not remain as they are."

ATTRIBUTED TO ST. AUGUSTINE

"What gives you hope?"

I hear this question from senior citizens worried about the world they're leaving their grandchildren, and from young moms wondering whether they should have brought a new life into this world. I hear it from fellow scientists, frustrated as their message falls on deaf ears, and from activists, worn out from years of advocacy with few visible results. I hear it from nearly anyone who reads the news these days, because the headlines do not give us hope: nearly everywhere we look, climate is changing faster or to a greater extent than previously thought. Ice sheets are melting. Sea level is rising. There are stronger hurricanes, out of control wildfires, record-breaking droughts—and don't even get me started on the politicization of basic facts and our seeming inability to treat with respect anyone we disagree with. Is it possible to find hope, in the midst of all of this?

———

Rick Lindroth is an ecologist at the University of Wisconsin in Madison. We met in the early 2000s, when we were both working on an assessment of climate change impacts on ecosystems in the Great Lakes region. Since then, I've gotten to know him much better, spoken at his church, which is similar to the one I attend, and shared concerns and frustrations regarding the state of our fellow believers, the science, and the world.

It's easy to see that Rick and his family don't live a life of conspicuous consumption. "We have what I call the 'Camry standard' for purchasing things," he says, referencing a reliable, mid-range model of car: "Get good quality items, but not those with all the bells and whistles." They live simply. "We were reducing our environmental footprint before people knew what a carbon footprint was," he says.

One of the most important things that people can do is to decide where they live in relation to where they work. Rick's family chose to live in a more expensive neighborhood, which has allowed him to walk or bike the two miles to his office year-round over the past thirty years. Until he and his wife had two teenage daughters, they did not buy a second car. They spent the better part of twenty-five years making their fifty-year-old home more energy-efficient as they could afford to do so—installing insulation, buying high-efficiency appliances, and putting in new windows. "We hardly ever use air-conditioning," says Rick. "We heat in the winter to 64°F during the day, and turn the furnace off at night. People who come to our house know to dress warm if it's winter or in shorts if it's summer."

As a scientist, he is not very hopeful about the way the climate is going. "I don't think that humankind will wake up sufficiently to the magnitude of the challenges that face us in a timely enough fashion to mitigate many of these enormous cataclysmic causes of suffering," he says.

At the same time, he is not altogether without hope. The rapid transformation of technology toward low carbon solutions, including renewable energy and low-carbon-emitting vehicles; the social activism that has escalated over the past year or two; and corporations that have independently stepped up and said, "Well, if the politicians aren't going to lead at least our company will"—all these give him some grounds for optimism.

He is also appreciative of and hopeful about the capacity of humankind—our nature, our creativity, our resilience, our ingenuity—to come up with solutions that are applicable and doable. "And I'm hopeful because of people like my kids," he says, "who are taking this seriously."

WHAT GIVES US HOPE

I've turned this question around and asked hundreds of people throughout the world what gives them hope: and it turns out, Rick is right. As humans, our hope is based on the idea of a future, and for most of us, the next generation embodies that future. I've asked people to explain what they mean by their answer, and most are very clear. All the kids taking action—through school strikes for climate, suing the federal government for the right to their future, winning science fairs for inventing algae biofuels and five-dollar water filters—are inspiring. But our hope isn't based on an expectation that they will fix it for us. Rather, we want to fix it for them. If there is no future, then who are we fighting to save the world for?

P. D. James's book *The Children of Men* chronicles the despair of a human race that can no longer bear children. Society has collapsed after a flu epidemic and plague has swept across the world. Refugee crises and hopelessness abound. As one of the characters says, "It was reasonable to struggle, to suffer, perhaps even to die, for a more just, a more compassionate society, but not in a world with no future where, all too soon, the very words 'justice,' 'compassion,' 'society,' 'struggle,' 'evil,' would be unheard echoes on an empty air."

In a beautiful essay called "The Concession to Climate Change I Will Not Make," Columbia Law School professor Jedediah Britton-Purdy shares his hope, the other side of this coin: to teach his son to marvel at the natural world before he realizes it is in peril. "When the thought of climate doom arrives, I hope it will arrive in a mind already prepared by curiosity and pleasure to know why this world is worth fighting to preserve," he writes. And eighteen-year-old Hannah Alper agrees. She's from Toronto, like me, and has been blogging about this issue since she was nine years old. "No matter how young you are, you can make a difference and you can be the change," she says.

WHAT HOPE IS NOT

There's a lot of false hope and fatalism out there, the idea that someone or something, nature or God or fate, will solve this problem for us without the need for human action. Both of these make us less likely to act, or to support others who do, because we feel like nothing matters.

I often hear this from other Christians. "God is in control," they say piously, "so shouldn't we just leave it in his hands?" Each time I wonder, haven't they read what the Bible says about reaping what you sow? God never promised to rescue us humans from the results of our bad decisions. Quite the opposite: the book of *Proverbs* warns, "Whoever sows injustice will reap calamity," and Hosea says, even more to the point, "for they have sown the wind, and they shall reap the whirlwind." Consequences aren't a punishment for sin, as some televangelists hasten to claim every time a disaster strikes. They're the simple consequence of the fact that we are all subject to the rules of physics. If humans increase heat-trapping gases in the atmosphere, the planet warms. Pretending we can defy physics by putting our heads in the sand or cultivating a positive attitude will merely keep us slightly happier until (and more surprised when) the axe falls.

Complacency and misplaced optimism are another form of false hope, and it's a bias that we humans are particularly vulnerable to, no matter who we are or where we live. Research has found we especially underestimate unfamiliar risks (and climate change is definitely that!) and optimistically assume we have more control over circumstances than we typically do. As Tali Sharot explains in her book *The Optimism Bias*, "we expect things to turn out better than they wind up being. People hugely underestimate their chances of getting divorced, losing their job or being diagnosed with cancer; envision themselves achieving more than their peers; and overestimate their likely life span (sometimes by 20 years or more)."

False hopes spring from our defense mechanisms. We employ them to deny and distract ourselves from a problem we feel helpless to face, or bad news we don't want to hear, like delaying going to the doctor when we anticipate a negative diagnosis. But while these false hopes might ease our mind short-term, they do nothing to erase the fears that still roil in the back of our brains.

That's why true hope must begin by recognizing the risk and understanding what's at stake. Rational hope accepts that success is not inevitable, or even entirely probable. It takes courage to do that, but when we are doubtful, when the odds are low and success is possible rather than probable, it's that courage and hope that carry us forward. Real hope also provides a vision of a future that we want to live in, where energy is abundant and available to all, where the economy is stable, where we have the resources we need, where our lives are not worse but better than they are today. It's a hope that is aware of all the others who are already working to make that future happen, and a hope that understands why we're doing this.

"We're not fighting for a merely 'livable' planet," says my friend and fellow climate scientist Peter Kalmus. "We're fighting for a riotous, wild, gorgeous, generous, miraculous, life-cradling planet that's home to a society that works for everyone."

That's what we are *all* looking for, and we are not alone.

WHERE MY HOPE COMES FROM

Real hope doesn't usually come knocking on the door of our brains uninvited, though. If we want to find it, we have to roll up our sleeves and go out and look for it. If we do, chances are we'll find it. And then we have to *practice* it.

The idea of hope as a practice, rather than an emotion or a value, has ancient roots in Buddhist philosophy. In their book *Active Hope: How to*

Face the Mess We're in Without Going Crazy, philosopher Joanna Macy and psychologist Chris Johnstone write:

> Active Hope is a practice. Like tai chi or gardening, it is something we do rather than have.... First, we take in a clear view of reality; second, we identify what we hope for; ... and third, we take steps to move ourselves or our situation in that direction.... Rather than weighing our chances and proceeding only when we feel hopeful, we focus on our intention and let it be our guide.

So that's what I do. I make a practice of hope. I search for and collect and share stories and good news about people who are making a difference, about the tech innovations like solar fabric, floating solar farms on flooded open-pit coal mines in China, river-powered energy in remote Arctic villages, and more. I participate in events and partner with organizations that share my values and promote advocacy and action—from museums to teachers' programs to faith-based initiatives. I offer them what I have: it might be my expertise, or my time, a donation or a skill. And I look to my faith; because the provenance for hope, the apostle Paul says, is not where we might think. It's not in rosy circumstances and positive conditions. It doesn't arrive when everything is going our way. In his epistle to the Romans, he lays it out: "We know that troubles help us learn not to give up. When we have learned not to give up, it shows we have stood the test. When we have stood the test, it gives us hope." Paul doesn't mention courage, but that runs through the entire verse. It takes courage to persevere under suffering; courage is part of character, and therefore courage is an essential ingredient of rational, constructive hope. And I love how he ends: "Hope never makes us ashamed because the love of God has come into our hearts." Love casts out fear, the Bible tells us, and it casts out shame as well. It's the glue that brings us together when every other force in the world, it seems, is trying to divide us into smaller and smaller tribes.

Science tells us it's too late to avoid all of the impacts of climate change. Some are already here today. Others are inevitable, because of the past choices we've made, and that can make us afraid. Science also tells us that much of what we do is actively contributing to the problem, from turning on our lights to what we eat for lunch. That makes us feel guilt. But the research I do is clear: it is *not* too late to avoid the most serious and dangerous impacts. Our choices will determine what happens.

The future we collectively face will be forged by our own actions. Climate change stands between us and a breathtaking, exhilarating future. We cannot afford to be paralyzed by fear or shame. We must act, with power, love, and a sound mind. Together, we can save ourselves.

ACKNOWLEDGMENTS

Giving a TED Talk is stressful. It's even worse when you somehow miss the fact that you have to memorize the entire twenty-minute talk, every single word of it. A lot of cramming and a considerable amount of hard kombucha went into that last week—and when I finished, the relief was indescribable. Heading for lunch, feeling like a thousand-pound weight had just rolled off my shoulders, I heard someone calling my name. "That was a great talk," a woman said. "Do you want to write a book?"

Flush with euphoria I thought, "Why not?" And that was how this book began.

Thanks first and foremost to that woman, whom I now know as my wonderful editor, Julia Cheiffetz, and to her capable associates Nicholas Ciani and Amara Balan at One Signal, for their inspiration to frame, write, and title this book. I'm also grateful to David Biello, who convinced me to give a TED talk in the first place and firmly rejected my proposal that it be about climate science. None of this would exist without them.

Thank you also to Sonia Smith and Georgina Ferry, who provided critical and in-depth feedback on the manuscript in its various forms. I first met Sonia many years ago when she was assigned to write a profile about me for *Texas Monthly*. An accomplished journalist, she's an absolute genius at nosing out the interesting people, places, and technologies that inspire some of the stories I tell in my PBS digital series,

Global Weirding, and in this book. I met Georgina much more recently, when *Nature* assigned her to edit the obituary I wrote for pioneering climate scientist Sir John Houghton, whose life was lost to COVID-19 in April 2020. Georgina's comments on the obituary were so masterly that I begged her to give the manuscript for this book a read. Being a lovely person, she did, and being a Brit, she called me on anything I said that would not be familiar or relevant to people outside of North American culture—so if you're part of that group, you have her to thank for making the book much more readable!

I also owe a huge debt of gratitude to my Texas Tech colleagues Bryan Giemza, Travis Snyder, and Ian Scott-Fleming, and my sister, Christy Hayhoe, who each took on the heroic task of reading the manuscript and providing invaluable critiques and insights and advice—and, in the case of Bryan, who is a literature professor, several good quotes as well.

As a scientist I'm accustomed to having everything I write peer-reviewed, so I am particularly grateful to my wonderful colleagues who provided some of that review for this manuscript. Their feedback was invaluable on topics that are not my own area of expertise—economics, social science, theology, and more—feedback that, in some cases, led to the dismantling and rewriting of entire chapters. These generous colleagues include theologian Chris Doran from Pepperdine University, whose book, *Hope in the Age of Climate Change*, reflects the foundation of faith that motivates my own work; ecologist-turned-social-scientist Jenn Marlon, from the Yale Program on Climate Communication, and health communication researcher Ed Maibach, from George Mason University, whose detailed data on public opinion inspired the title of my TED Talk and so much of the discussion in the first third of the book; communication experts George Marshall and Leane de Laigue from Climate Outreach, whose research and wisdom never fails to inspire me; social scientist Brandi Morris from Aarhus University, who serendipitously emailed me at the very moment I was struggling with incorporating her challenging research into the chapter on fear, and got roped into reviewing it; economists

Gernot Wagner from New York University, Max Auffhammer from the University of California at Berkeley, and Andrew Leach from the University of Alberta, who provided much appreciated insight into the finer nuances of carbon pricing and cap and trade that expanded my own horizons; my new colleague David Banks from The Nature Conservancy, who's been giving me a crash course on nature-based solutions; sociologist Kari Norgaard from the University of Oregon, who untangles the complicated reasons we don't act on climate; Ed Hawkins from the University of Reading, fellow climate scientist and creator of the Warming Stripes and many other creative and compelling climate graphics; permafrost experts Ted Schuur from Northern Arizona University and Katey Anthony from the University of Alaska, who have spent so much of their lives in the most remote areas of the world, trying to figure out what Arctic warming means for our world; ecologist Terry Chapin, whose invitation to Alaska inspired me and whose thoughtful leadership has done so much for the scientific community; David Fenton, pioneering public relations expert and tireless climate advocate, who encourages and informs my work; and solar expert Andreas Karelas from RE-volv, an organization that helps nonprofits invest in clean energy, who wrote the inspirational book *Climate Courage: How Tackling Climate Change Can Build Community, Transform the Economy, and Bridge the Divide in America*, for which I provided the foreword. I am also deeply appreciative of the hundreds of studies and dozens of books that provide a deep dive into so many of the topics I tackle here, many of which I cite in the chapter notes. I stand on their wisdom and learning and take full responsibility for any errors of interpretation or fact.

While facts are important, though, it's stories, so neuroscientists tell us, that are essential to communication. As I mentioned, stories literally synchronize our brain waves as we empathize with one another—and when it comes to climate change, there's a lot to empathize with. So last but perhaps most profoundly, I am incredibly grateful to all whose stories I share here, including those who have spoken with me over the years and in particular John Cook, Simon Donner, David Folland, Tim

Fullman, Glyn Goodwin, Don Lieber, Rick Lindroth, Kirstin Milks, and Renée Rostius, who offered their experiences specifically for this book. Without their stories and those of so many others, this book would be a lot less interesting to read. And honestly, without all of these stories there probably wouldn't even be a book: because it was hearing from people like Glyn that convinced me that it was even possible for this to make a difference.

I hope you'll agree—and if you do, I'd love to hear *your* story.

NOTES

Part of chapter 8 was previously published in *Foreign Policy* as "Yeah, the Weather Has Been Weird" in May 2017; two paragraphs in chapter 14 were previously published in the PLOS SciComm blog as "The climate is changing. Why does that matter to me and why should it matter to you?" in March 2019; and a paragraph in chapter 2 was previously published in the *New York Times* as "I'm a Climate Scientist Who Believes in God. Hear Me Out" in October 2019.

PREFACE

x *The U.S. is arguably home* Michael Dimock and Richard Wike, "America Is Exceptional in the Nature of Its Political Divide," Pew Research Center, November 23, 2010, https://www.pewresearch.org/fact-tank/2020/11/13/america-is-exceptional-in-the-nature-of-its-political-divide/.

x *The Beyond Conflict Institute's 2020 report* Beyond Conflict, *America's Divided Mind*, June 2020, https://beyondconflictint.org/wp-content/uploads/2020/06/Beyond-Conflict-America_s-Div-ided-Mind-JUNE-2020-FOR-WEB.pdf.

x *climate change tops that list* Pew Research Center, "Election 2020: Voters Are Highly Engaged, but Nearly Half Expect to Have Difficulties Voting," August 2020, https://www.pewresearch.org/politics/2020/08/13/election-2020-voters-are-highly-engaged-but-nearly-half-expect-to-have-difficulties-voting/.

x *Over 50 percent* Anthony Leiserowitz, Edward Maibach, Connie Roser-Renouf, Seth Rosenthal, and Teresa Myers, Global Warming's Six Ameri-

cas, Yale Program on Climate Communication, https://climatecommu
nication.yale.edu/about/projects/global-warmings-six-americas/.

x *"there is a strong sense"* Matthew Smith, "International Poll: Most Expect to Feel Impact of Climate Change, Many Think It Will Make Us Extinct," YouGov, December 14, 2019, https://yougov.co.uk/topics/science
/articles-reports/2019/09/15/international-poll-most-expect-feel-impact
-climate.

x *In the U.S.* American Psychological Association, "Majority of US Adults
Believe Climate Change Is Most Important Issue Today," February 6,
2020, https://www.apa.org/news/press/releases/2020/02/climate-change.

xii *are acting according to them* Jonathan Haidt, *The Righteous Mind: Why
Good People Are Divided by Politics and Religion* (New York: Random
House, 2012).

SECTION 1:
THE PROBLEM AND THE SOLUTION

CHAPTER 1

3 *a common folk theory* George Lakoff, *Don't Think of An Elephant!* (White
River, VT: Chelsea Green Publishing, 2014).

5 *In 1998, a Gallup poll found* Gallup, "Partisan Gap on Global Warming
Grows," May 29, 2008, https://news.gallup.com/poll/107593/partisan-gap
-global-warming-grows.aspx.

5 *As recently as 2008* Michael O'Brien, "Gingrich Regrets 2008 Climate Ad
with Pelosi," *The Hill,* July 26, 2011, https://thehill.com/blogs/blog-brief
ing-room/news/173463-gingrich-says-he-regrets-2008-climate-ad-with
-pelosi.

5 *By 2020, coronavirus* Pew Research Center, "Election 2020: Voters Are
Highly Engaged, but Nearly Half Expect to Have Difficulties Voting,"
August 2020, https://www.pewresearch.org/politics/2020/08/13/election
-2020-voters-are-highly-engaged-but-nearly-half-expect-to-have-difficul
ties-voting/.

5 *In Canada* Yale Program on Climate Communication, "Canadian Climate Opinion Maps 2018," November 21, 2019, https://climatecommu
nication.yale.edu/visualizations-data/ccom/.

5 *Conservative members of parliament* Jonathan Watts and Pamela Duncan,
"Tory MPs Five Times as Likely to Vote Against Climate Action," *Guardian,* October 11, 2019, https://www.theguardian.com/environment/2019
oct/11/tory-mps-five-times-more-likely-to-vote-against-climate-action.

6 *Their claims were bolstered by disinformation* Timothy Graham and To-

bias Keller, "Bushfires, Bots and Arson Claims: Australia Flung in The Global Disinformation Spotlight," *The Conversation*, January 10, 2020, https://theconversation.com/bushfires-bots-and-arson-claims-australia-flung-in-the-global-disinformation-spotlight-129556.

6 *people's opinions on climate change* Matthew Hornsey, Emily Harris, Paul Bain, and Kelly Fielding, "Meta-Analyses of the Determinants and Outcomes of Belief in Climate Change," *Nature Climate Change* 6, no. 6 (2016): 622–626, https://doi.org/10.1038/nclimate2943.

6 *the polarization is emotional* Greg Lukianoff and Jonathan Haidt, *The Coddling of the American Mind: How Good Intentions and Bad Ideas Are Setting Up a Generation for Failure* (London: Penguin Books/Random House U.K., 2018).

6 *many immediately changed their opinions* Cass Sunstein, *How Change Happens* (Cambridge: MIT Press, 2019).

6 *"As media frames opposing viewpoints"* Tania Israel, *Beyond Your Bubble: How to Connect Across the Political Divide* (Washington, DC: APA Life-Tools, 2020).

7 *Senator Ted Cruz told Glenn Beck* Glenn Beck, https://www.glennbeck.com/2015/10/29/ted-cruz-climate-change-is-not-science-its-religion/.

7 *Senator Lindsey Graham said* Council on Foreign Relations, https://www.cfr.org/event/lindsey-graham-irans-nuclear-program.

8 *Called* Global Warming's Six Americas Anthony Leiserowitz, Edward Maibach, Connie Roser-Renouf, Seth Rosenthal, and Teresa Myers, *Global Warming's Six Americas*, Yale Program on Climate Communication, https://climatecommunication.yale.edu/about/projects/global-warmings-six-americas/, accessed September 2020.

9 *the helpful Skeptical Science website* Skeptical Science, https://skepticalscience.com/.

9 *cognitive linguist George Lakoff explains* George Lakoff, *Don't Think of An Elephant!* (White River, VT: Chelsea Green Publishing, 2014).

10 *Research on everything from airplane seatbelts* Tali Sharot, *The Influential Mind: What the Brain Reveals About Our Power to Change Others* (New York: Henry Holt and Co., 2017).

CHAPTER 2

13 *"Climate change public communication"* Anthony Leiserowitz, Edward Maibach, Connie Roser-Renouf, Seth Rosenthal, and Teresa Myers, *Global Warming's Six Americas*, Yale Program on Climate Communication, https://climatecommunication.yale.edu/about/projects/global-warmings-six-americas/, accessed September 2020.

14 *26,500 independent lines of evidence* Cynthia Rosenzweig, David Karoly, Marta Vicarelli, Peter Neofotis, Qigang Wu, Gino Casassa, Annette Menzel, et al. "Attributing physical and biological impacts to anthropogenic climate change," *Nature* 453 (2008): 353–357, https://doi.org/10.1038/nature06937.

14 *Fort Hood now draws* American Association for the Advancement of Science, How We Respond: Community Responses to Climate Change (2019), https://howwerespond.aaas.org/community-spotlight/fort-hood-embraces-renewable-energy-other-military-posts-follow-suit/.

14 *As of 2020* American Wind Energy Association, State Fact Sheets, https://www.awea.org/resources/fact-sheets/state-facts-sheets; Solar Energy Industries Association, States Map, https://www.seia.org/states-map.

16 *a study I did for* **Sports Illustrated** *Sports Illustrated* staff, "Going, Going Green," *Sports Illustrated*, March 12, 2007, https://vault.si.com/vault/2007/03/12/going-going-green.

17 *as many as a billion* Institute for Economics & Peace, Ecological Threat Register, 2020, https://ecologicalthreatregister.org/.

18 *white evangelicals are less worried* Frank Kummer, "Religious people believe climate change is a real threat, not a controversy, Yale poll finds," *The Philadelphia Inquirer*, October 23, 2020, https://www.inquirer.com/science/climate/climate-change-religion-amy-coney-20201023.html.

18 *it's the political polarization* John H. Evans and Justin Feng, "Conservative Protestantism and skepticism of scientists studying climate change," *Climatic Change* 121 (2013): 595–608, https://doi.org/10.1007/s10584-013-0946-6.

19 *study after study has shown* Emily Kubin, Curtis Puryear, Chelsea Schein, and Kurt Gray, "Personal experiences bridge moral and political divides better than facts," *Proceedings of the National Academy of Science of the United States* 118, no. 6 (2021): e2008389118, https://doi.org/10.1073/pnas.2008389118.

CHAPTER 3

22 *calls it a* **threat multiplier** Department of Defense, 2014 Climate Change Adaptation Roadmap, https://www.acq.osd.mil/eie/Downloads/CCAR print_wForward_e.pdf.

26 *The Audubon Society has put together* The National Audubon Society, "Survival by Degrees: 389 Bird Species on the Brink," 2020, https://www.audubon.org/climate/survivalbydegrees.

26 *In Oregon and Idaho, it's estimated* Pete Bisson, "Salmon and Trout in the Pacific Northwest and Climate Change," June 2008, U.S. Department of

Agriculture, Forest Service, Climate Change Resource Center, www.fs.fed
.us/ccrc/topics/aquatic-ecosystems/salmon-trout.shtml.

26 *Ducks Unlimited says that conservation* Ducks Unlimited, "Climate
Change and Waterfowl, 2020," https://www.ducks.org/Conservation
/Public-Policy/Climate-Change-and-Waterfowl.

26 *a study I led for the U.S. Northeast* Peter Frumhoff, James McCarthy,
Jerry Melillo, Suzanne Moser, and Donald Wuebbles, *Confronting Climate
Change in the U.S. Northeast: Science, Impacts, and Solutions: Synthesis
Report of the Northeast Climate Impacts Assessment (NECIA)* (Cambridge,
MA: Union of Concerned Scientists, 2007).

27 *affecting the viability of many other outdoor sports* Priestley International Centre for Climate, "Game Changer: How Climate Change Is Impacting Sports in the U.K.," 2018, https://climate.leeds.ac.uk/wp-content
/uploads/2018/02/Game-Changer-1.pdf.

27 *the Texas Rangers baseball team* Justin Fox, "It's Gotten Too Hot for
Outdoor Baseball in Texas," *Bloomberg Opinion*, 2019, https://www
.bloomberg.com/opinion/articles/2019-10-01/climate-change-ruined
-globe-life-stadium-for-the-texas-rangers.

27 *Outdoor ice rinks* Nikolay Damyanov, Damon Matthews, and Lawrence
Mysak, "Observed Decreases in the Canadian Outdoor Skating Season
Due to Recent Winter Warming," *Environmental Research Letters* 7, no. 1
(2012): 014028, https://doi.org/10.1088/1748-9326/7/1/014028.

27 *Summer Olympic hosts like Tokyo* Katherine Kornei, "Japan: Next Olympics Marathon Course Has Dangerous 'Hot Spots' for Spectators," *Prevention Web: The Knowledge Platform for Disaster Risk Reduction*, 2019,
https://www.preventionweb.net/news/view/63205.

27 *Outdoor tennis competitions* Graham Readfern, "Is the Australian Open
Tennis Feeling the Heat of Climate Change?," *Guardian*, January 16,
2014, https://www.theguardian.com/environment/planet-oz/2014/jan/16
/australia-tennis-open-climate-change-extreme-heat.

27 *Half the world's sandy beaches* Michalis Vousdoukas, Roshanka Ranas-
inghe, Lorenzo Mentaschi, Theocharis Plomaritis, Panagiotis Athana-
siou, Arjen Luijendijk, and Luc Feyen, "Sandy Coastlines Under Threat
of Erosion," *Nature Climate Change* 10 (2020): 260–263, https://doi
.org/10.1038/s41558-020-0697-0.

27 *plant hardiness zones shifting* Arbor Day Foundation, "Zone Changes:
1990 USDA Hardiness Zones," 2020, https://www.arborday.org/media
/mapchanges.cfm.

27 *over a trillion dollars since 1970* Christophe Diagne, Boris Leroy, Anne-
Charlotte Vaissière, Rodolphe Gozlan, David Roiz, Ivan Jarić, Jean-Michel
Salles, Corey Bradshaw, and Franck Courchamp, "High and rising eco-

nomic costs of biological invasions worldwide," *Nature* 592 (2021): 571–576, https://doi.org/10.1038/s41586-021-03405-6.

28 *created brochures and a website* Bethany Bradley, Amanda Bayer, Bridget Griffin, Sydni Joubran, Brittany Laginhas, et al., "Gardening with Climate-Smart Native Plants in the Northeast, 2020" https://scholar works.umass.edu/eco_ed_materials/8/ and https://www.risccnetwork .org/.

28 *reached southern Ontario* Ontario Invasive Plant Council, Kudzu, https:// www.ontarioinvasiveplants.ca/wp-content/uploads/2016/07/Kudzu -ENGLISH.pdf.

28 *Greens like lettuce* Maryn McKenna, "Can Lettuce Survive Climate Change?" Wired, February 7, 2019, https://www.wired.com/story/can -lettuce-survive-climate-change/.

28 *water supplies needed* Varun Varma and Daniel Bebber, "Climate change impacts on banana yields around the world," *Nature Climate Change* (2019) doi: 10.1038/s41558-019-0559-9.

28 *Garden Club of America* Garden Club of America, "Position Paper: Climate Change," 2016, https://www.gcamerica.org/public/assets/pdf/GCA PositionPapersCompilation2016.pdf.

28 *destroy between 20 to 40 percent* Food and Agriculture Organization of the United Nations, "Plants vital to human diets but face growing risks from pests and diseases," April 4, 2016, http://www.fao.org/news/story /en/item/409158/icode/.

28 *wine-growing areas of France* Cornelis van Leeuwen and Philippe Darriet, "The Impact of Climate Change on Viticulture and Wine Quality," *Journal of Wine Economics* 11, no. 1: 150–167.

28 *decreased the yield of hops* Martin Mozny, Radim Tolasz, Jiri Nekovar, Tim Sparks, Mirek Trnka, and Zdenek Zalud, "The Impact of Climate Change on the Yield and Quality of Saaz Hops in the Czech Republic," *Agricultural and Forest Meteorology* 149, no. 6–7 (2009): 913–919, https:// doi.org/10.1016/j.agrformet.2009.02.006.

28 *Brewery Climate Declaration* Caitlyn Kennedy, "Climate and Beer," NOAA Climate.gov, 2016, https://www.climate.gov/news-features/cli mate-and/climate-beer.

28 *cassava root as a replacement for barley* Andrea Shea, "Survival of the Greenest Beer? Breweries Adapt to a Changing Climate," *NPR: The Salt*, 2015, https://www.npr.org/sections/thesalt/2015/06/24/415538451/sur vival-of-the-greenest-beer-breweries-adapt-to-a-changing-climate.

29 *Shifts in rainfall patterns* Michon Scott, "Climate and Chocolate," NOAA Climate.gov, 2016, https://www.climate.gov/news-features/climate-and /climate-chocolate.

29 *squeezing the water out of the soil* The International Center for Tropical Agriculture (CIAT), "CIAT Warns of Climate Change Impact on Cocoa Production," *IISD/SDG Knowledge Hub*, 2011, https://sdg.iisd.org/news /ciat-warns-of-climate-change-impact-on-cocoa-production/.

29 *Coffee giants from Nespresso to Lavazza* Lavazza Foundation, "Coffee and Climate: Changing the Future One Cup at a Time," *Guardian Labs*, https://www.theguardian.com/lavazza-ethical-espresso/2019/feb/26 /coffee-and-climate-changing-the-future-one-cup-at-a-time.

29 *programs to build resilience* Karen Carmichael, "Easing the Impact of Climate Change on Coffee Growers," *National Geographic*, 2020, https:// www.nationalgeographic.com/science/2020/02/partner-content-impact -climate-change-on-coffee-growers/.

30 *Warming stripes are* #showyourstripes, https://showyourstripes.info/.

SECTION 2:
WHY FACTS MATTER—AND WHY THEY'RE NOT ENOUGH

CHAPTER 4

39 *analysis of people across fifty-six countries* Matthew Hornsey, Emily Harris, Paul Bain, and Kelly Fielding, "Meta-Analyses of the Determinants and Outcomes of Belief in Climate Change," *Nature Climate Change* 6, no. 6 (2016): 622–626, https://doi.org/10.1038/nclimate2943.

40 *Colbert tweeted sarcastically* Stephen Colbert, Twitter post, November 18, 2014, 11:40 p.m., https://twitter.com/StephenAtHome/status/534929076 726009856.

40 *mathematician and scientist Joseph Fourier* Joseph Fourier, "*Remarques generales sur les temperatures du globe terrestre et des espaces planetaires*" (1824), *Annales de chimie et de physique* (*Annals of Chemistry and of Physics*) 23, no. 2 (1999).

40 *a paper presented on her behalf* Eunice Newton Foote, "Circumstances Affecting the Heat of the Sun's Rays," Annual Meeting of the American Association for the Advancement of Science, August 23, 1856.

40 *needed to measure precisely* John Tyndall, "On the Absorption and Radiation of Heat by Gases and Vapours . . ." *Philosophical Magazine* 4, 22 (1861): 169–194.

41 *Guy Callendar could actually measure* G. S. Callendar, "The Artificial Production of Carbon Dioxide and Its Influence on Temperature," *Quarterly Journal of the Royal Meteorological Society* 64, no. 275 (1938): 223–240.

41 *"Within a few short centuries"* Ad Hoc Study Group on Carbon Dioxide and Climate, *Carbon Dioxide and Climate: A Scientific Assessment*, Climate

Research Board, Assembly of Mathematical and Physical Sciences, National Research Council (Washington D.C.: National Academy of Sciences, 1979).

41 *"prevent dangerous anthropogenic"* United Nations Framework Convention on Climate Change, 1992, https://unfccc.int/resource/docs/convkp/conveng.pdf.

41 *known as the Paris Agreement* United Nations Paris Agreement, 2015, https://unfccc.int/files/meetings/paris_nov_2015/application/pdf/paris_agreement_english_.pdf.

41 *"We, as leaders of major scientific organizations"* Scientific Societies' Letter to Congress, June 28, 2016, https://www.eurekalert.org/images/2016climateletter6-28-16.pdf.

42 *one hundred and ninety-eight scientific organizations* https://www.opr.ca.gov/facts/list-of-scientific-organizations.html.

43 *In the Northern Hemisphere* Drew Shindell, Gavin Schmidt, Michael Mann, David Rind, and Anne Waple, "Solar Forcing of Regional Climate Change during the Maunder Minimum," *Science* 294, no. 5549 (2001): 2149–2152, https://doi.org/10.1126/science.1064363.

43 *satellite radiometer data show* Chi Ju Wu, Natalie Krivova, Sami Solanki, and Ilya Usoskin, "Solar Total and Spectral Irradiance Reconstruction Over the Last 9,000 Years," *Astronomy & Astrophysics* 620 (2018): A120, https://doi.org/10.1051/0004-6361/201832956.

43 *over 60 megatons* Achmad Djumarma Wirakusumah and Heryadi Rachmat, "Impact of the 1815 Tambora Eruption to Global Climate Change," *IOP Conference Series: Earth and Environmental Science* 71 (2016): 012007.

44 *outbreaks of typhus* Clive Oppenheimer, "Climatic, Environmental and Human Consequences of the Largest Known Historic Eruption: Tambora Volcano (Indonesia) 1815," *Progress in Physical Geography* 27, no. 2 (2003): 230–259, https://doi.org/10.1191%2F0309133303pp379ra.

44 *ultimately responsible for* Matthew Genge, "Electrostatic Levitation of Volcanic Ash into the Ionosphere and Its Abrupt Effect on Climate," *Geology* 46, no. 10 (2018): 835–838, https://doi.org/10.1130/G45092.1.

44 *natural geologic emissions* Terry Gerlach, "Volcanic Versus Anthropogenic Carbon Dioxide," *EOS Science News* 92, no. 24 (2011): 201–202, https://doi.org/10.1029/2011EO240001.

44 *Milutin Milanković figured out* Milutin Milanković, *Canon of Insolation and the Ice-age Problem (Kanon Der Erdbestrahlung und Seine Anwendung Auf Das Eiszeitenproblem)*, 4th ed. (Agency for Textbooks, 1941).

45 *in about fifteen hundred years* Chronis Tzedakis, James Channell, David Hodell, Helga Kleiven, and Luke Skinner, "Determining the Natural Length of the Current Interglacial," *Nature Geoscience* 5 (2012): 138–141, https://doi.org/10.1038/ngeo1358.

45 *It also typically brings* Rebecca Lindsey, "Global impacts of El Niño and La Niña," NOAA Climate.gov, February 9, 2016, https://www.climate .gov/news-features/featured-images/global-impacts-el-ni%C3%B1o-and -la-ni%C3%B1a.

46 *so-called Medieval Warm Period* Eystein Jansen, Jonathan Overpeck, Keith Briffa, Jean-Claude Duplessy, Fortunat Joos, Valerie Masson-Delmotte, Daniel Olago, et al., "Palaeoclimate," in *Climate Change 2007: The Physical Science Basis. Contribution of Working Group I to the Fourth Assessment Report of the Intergovernmental Panel on Climate Change*, eds. Susan Solomon, Dahe Qin, Martin Manning, Zhenlin Chen, Melinda Marquis, Kristen Averyt, Melinda Tignor, and Henry LeRoy Miller (Cambridge and New York: Cambridge University Press, 2007).

46 *ocean heat content* John Abraham, Molly Baringer, Nathan Bindoff, Tim Boyer, Lijing Cheng, John Church, Jessica Conroy, et al., "A Review of Global Ocean Temperature Observations: Implications for Ocean Heat Content Estimates and Climate Change," *Reviews of Geophysics* 51, no. 3 (2013): 450–483, https://doi.org/10.1002/rog.20022.

47 *more than 100 percent* Thomas Knutson, James Kossin, Carl Mears, Judith Perlwitz, and Michael Wehner, "Detection and Attribution of Climate Change," in *Climate Science Special Report: Fourth National Climate Assessment, Volume I*, eds. Donald Wuebbles, David Fahey, Kathy Hibbard, David Dokken, Brooke Stewart, and Thomas Maycock (Washington, DC: U.S. Global Change Research Program, 2018), http://doi.org/10.7930 /J01834ND.

47 *enough to stave off* William Ruddiman, *Plows, Plagues, and Petroleum: How Humans Took Control of Climate*, 2nd ed. (Princeton: Princeton University Press, 2016).

47 *more than 420 ppm* Earth System Research Laboratories Global Monitoring Laboratory, "Carbon Cycle Greenhouse Gases: Trends in CO_2," https://gml.noaa.gov/ccgg/trends/.

47 *15 million years ago* Aradhna Tripati, Christopher Roberts, and Robert Eagle, "Coupling of CO2 and Ice Sheet Stability Over Major Climate Transitions of the Last 20 Million Years," *Science* 326, no. 5958 (2009): 1394–1397, http://doi.org/10.1126/science.1178296.

47 *55 million years* Katharine Hayhoe, James Edmonds, Robert Kopp, Allegra LeGrande, Benjamin Sanderson, Michael Wehner, and Donald Wuebbles, "Climate Models, Scenarios, and Projections," in *Climate Science Special Report: Fourth National Climate Assessment, Volume I*, eds. Donald Wuebbles, David Fahey, Kathy Hibbard, David Dokken, Brooke Stewart, and Thomas Maycock (Washington, DC: U.S. Global Change Research Program, 2018), http://doi.org/10.7930/J0WH2N54.

47 *sea level was over* Ibid.

47 *ten times the pace* Richard Zeebe, Andy Ridgwell, and James Zachos, "Anthropogenic Carbon Release Rate Unprecedented During the Past 66 Million Years," *Nature Geoscience* 9 (2016): 325–329, http://doi.org/10.1038/ngeo2681.

CHAPTER 5

49 *We are so locked* Ezra Klein, *Why We're Polarized* (New York: Avid Reader Press / Simon & Schuster Inc., 2020).

49 *2005 hurricane season* National Weather Service, "2005 Hurricane Season Records," https://www.weather.gov/tae/climate_2005review_hurricanes.

50 *most expensive tropical storm* Ibid.

51 *There's even a book about it* James Hoggan, *I'm Right and You're an Idiot: The Toxic State of Public Discourse and How to Clean it Up*, 2nd ed. (Gabriola Island, BC: New Society Publishers, 2019).

52 *"ordinary science intelligence"* Dan Kahan, " 'Ordinary Science Intelligence': A Science-Comprehension Measure for Study of Risk and Science Communication with Notes on Evolution and Climate Change," *Journal of Risk Research* 20, no. 8 (2016): 995–1016, https://doi.org/10.1080/13669877.2016.1148067.

52 *People who score high* Anthony Leiserowitz, N. Smith, and Jennifer Marlon, "Americans' Knowledge of Climate Change," (New Haven: Yale University Project on Climate Change Communication, 2010).

53 *"people with the highest degree"* Dan Kahan, Ellen Peters, Maggie Wittlin, Paul Slovic, Lisa Larrimore Ouellette, Donald Braman, and Gregory Mandel, "The Polarizing Impact of Science Literacy and Numeracy on Perceived Climate Change Risks," *Nature Climate Change* 2 (2012): 732–735, https://doi.org/10.1038/nclimate1547.

53 *a more recent study* Gabriela Czarnek, Malgorzata Kossowska, and Paulina Szwed, "Right-Wing Ideology Reduce the Effects of Education on Climate Change Beliefs in More Developed Countries," *Nature Climate Change* 11 (2021): 9–13, https://doi.org/10.1038/s41558-020-00930-6.

53 *we ask ourselves* Thomas Gilovich, *How We Know What Isn't So: The Fallibility of Human Reason in Everyday Life* (New York, NY: The Free Press, 1991).

54 *this type of motivated reasoning* D. Perkins, "Postprimary Education Has Little Impact on Informal Reasoning," *Journal of Educational Psychology* 77, no. 5 (1985): 562–571, https://psycnet.apa.org/doi/10.1037/0022-0663.77.5.562.

55 *Rasmus Benestad, collected the studies* Rasmus Benestad, Dana Nuccitelli, Stephan Lewandowsky, Katharine Hayhoe, Hans Olav Hygen, Rob van Dorland, and John Cook, "Learning from Mistakes in Climate Research," *Theoretical and Applied Climatology* 126 (2016): 699–703, https://doi.org/10.1007/s00704-015-1597-5.

55 *As Peter Boghossian and James Lindsay explain* Peter Boghossian and James Lindsay, *How to Have Impossible Conversations* (New York: Lifelong Books, 2019).

56 *a personal attack on their identity* Stephan Lewandowsky, Ullrich Ecker, Colleen Seifert, Norbert Schwarz, and John Cook, "Misinformation and Its Correction: Continued Influence and Success," *Psychological Science in the Public Interest* 13, no. 3 (2012): 106–131, https://doi.org/10.1177%2F1529100612451018.

56 *a kind of* **backfire effect** Jack Zhou, "Boomerangs Versus Javelins: How Polarization Constrains Communication on Climate Change," *Environmental Politics* 25, no. 5 (2016): 788–811, https://doi.org/10.1080/09644016.2016.1166602.

58 *the value of belonging* Dan Kahan, Ellen Peters, Maggie Wittlin, Paul Slovic, Lisa Larrimore Ouellette, Donald Braman, and Gregory Mandel, "The Polarizing Impact of Science Literacy and Numeracy on Perceived Climate Change Risks," *Nature Climate Change* 2 (2012): 732–735, https://doi.org/10.1038/nclimate1547.

58 *one study where researchers* Christopher Bail, Lisa P. Argyle, Taylor W. Brown, John P. Bumpus, Haohan Chen, M. B. Fallin Hunzaker, Jaemin Lee, Marcus Mann, Friedolin Merhout, and Alexander Volfovsky, "Exposure to Opposing Views on Social Media Increase Political Polarization," *Proceedings of the National Academy of Sciences*, 115, no. 37 (2018): 9216–9221, https://doi.org/10.1073/pnas.1804840115.

58 *"get a kick out of information."* Tali Sharot, *The Influential Mind: What the Brain Reveals About Our Power to Change Others* (New York: Henry Holt and Co., 2017).

60 *teaching kids about climate change* Danielle Lawson, Kathryn Stevenson, Nils Peterson, Sarah Carrier, Renee Strand, and Peter Seekamp, "Children Can Foster Climate Change Concern Among Their Parents," *Nature Climate Change* 9 (2019): 458–462, https://doi.org/10.1038/s41558-019-0463-3.

CHAPTER 6

63 *"Climate change contains"* George Marshall, *Don't Even Think About It: Why Our Brains Are Wired to Ignore Climate Change* (New York: Bloomsbury USA, 2015).

63 *"Is the Earth f*cked?"* Eli Kintisch, "'Is Earth F**ked?' AGU Scientist Asks," *Science*, December 2012, https://www.sciencemag.org/news/2012/12/earth-fked-agu-scientist-asks.

63 *"a weary group of top climatologists"* "Sighing, Resigned Climate Scientists Say to Just Enjoy Next 20 Years as Much as You Can," *The Onion*, February 23, 2018, https://www.theonion.com/sighing-resigned-climate-scientists-say-to-just-enjoy-1823265249.

64 *"err on the side of least drama"* Keynyn Brysse, Naomi Oreskes, Jessica O'Reilly, and Michael Oppenheimer, "Climate change prediction: Erring on the side of least drama?," *Global Environmental Change* 23, no. 1 (2013): 327–337, https://doi.org/10.1016/j.gloenvcha.2012.10.008.

64 *studies have found* Robert Kopp, Katharine Hayhoe, David Easterling, Timothy Hall, Radley Horton, Kenneth Kunkel, and Allegra LeGrande, "Potential Surprises—Compound Extremes and Tipping Elements, in *Climate Science Special Report: Fourth National Climate Assessment, Volume I*, eds. Donald Wuebbles et al. (Washington, D.C.: U.S. Global Change Research Program, 2017), pp. 411–429, https://doi.org/10.7930/J0GB227J.

65 *"the systematic tendency of climate models"* Ibid.

65 *pessimistic messages increase risk perception* Brandi Morris, Polymeros Chrysochou, Simon Karg, and Panagiotis Mitkidis., "Optimistic vs. Pessimistic Endings in Climate Change Appeals," *Humanities and Social Sciences Communications* 7 (2020): 82, https://doi.org/10.1057/s41599-020-00574-z.

66 *fear works well when coupled* Tali Sharot, *The Influential Mind: What the Brain Reveals About Our Power to Change Others* (New York: Henry Holt and Co., 2017).

66 *there is a substantial emotional cost* Cass R. Sunstein, *Too Much Information: Understanding What You Don't Want to Know* (Cambridge, MA: The MIT Press, 2020).

67 *It was fear that motivated* David Wallace-Wells, *The Uninhabitable Earth: Life After Warming* (New York: Tim Duggan Books, 2019).

67 *"I was reading The Uninhabitable Earth"* Andy Revkin, "Thriving Online: Three Young Leaders Building Impact Networks for Sustainable Societies," pscp.tv, 2020, https://www.pscp.tv/Revkin/1YpKkNQoLqNxj?t=24m18s.

68 *Andreas talks about how he saw* Andreas Karelas, *Climate Courage: How Tackling Climate Change Can Build Community, Transform the*

Economy, and Bridge the Political Divide in America (Boston: Beacon Press, 2020).

69 *there's little evidence* Saffron O'Neill and Sophie Nicholson-Cole, "Fear Won't Do It: Promoting Positive Engagement with Climate Change Through Visual and Iconic Representations," *Science Communication* 30, no. 3 (2009), https://doi.org/10.1177/1075547008329201.

69 *fear-based messaging can trigger* Rebecca Huntley, *How to Talk About Climate Change in a Way That Makes a Difference* (Sydney: Murdoch Books, 2020).

69 *"We do not accept climate change"* George Marshall, *Don't Even Think About It: Why Our Brains Are Wired to Ignore Climate Change* (New York: Bloomsbury USA, 2015).

69 *"The human brain is built to associate"* Tali Sharot, *The Influential Mind: What the Brain Reveals About Our Power to Change Others* (New York: Henry Holt and Co., 2017).

70 *"When the body releases stress chemicals"* Katie Patrick, *How to Save the World: How to Make Changing the World the Greatest Game We've Ever Played* (San Francisco: Hello World Labs, 2020).

70 *she talks about the grief* Christiana Figueres and Tom Rivett-Carnac, *The Future We Choose: Surviving the Climate Crisis* (New York: Alfred A. Knopf, 2020).

CHAPTER 7

73 *"No one can unilaterally"* Leah Cardamore Stokes, "A Field Guide For Transformation," in *All We Can Save: Truth, Courage, and Solutions for the Climate Crisis* (New York: One World/Random House, 2020).

74 As *psychologist and marketing expert* Robert Cialdini, Influence, New and Expanded: The Power of Persuasion (New York: Harper Business, 2021).

75 *"flygskam," or "flight shame"* William Wilkes and Richard Weiss, "German Air Travel Slump Points to Spread of Flight Shame," *Bloomberg*, December 18, 2019, https://www.bloomberg.com/news/articles/2019-12-19/german -air-travel-slump-points-to-spread-of-flight-shame.

77 *when people were told to change* Risa Palm, Toby Bolsen, and Justin Kingsland, " 'Don't Tell Me What to Do': Resistance to Climate Change Messages Suggesting Behaviour Changes," *Weather, Climate and Society* 12, no. 4 (2020): 75–84, http://doi.org/10.1175/WCAS-D-19-0141.1.

77 *average of 2 percent* Hunt Allcott, "Social norms and energy conservation," *Journal of Public Economics* 95, no. 9–10 (2011): 1082–1095, https:// doi.org/10.1016/j.jpubeco.2011.03.003.

77 *over $1 billion in bills* Robert Walton, "Opower hits 11 TWh in energy

savings milestone," *Utility Dive*, June 13, 2016, https://www.utilitydive.com/news/opower-hits-11-twh-in-energy-savings-milestone/420787/.

77 *"households that were politically conservative"* P. W. Schultz, J. Nolan, R. Cialdini, N. Goldstein, and V. Griskevicius, "The Constructive, Destructive, and Reconstructive Power of Social Norms: Reprise," *Perspectives on Psychological Science* 13 (2018): 249–254, https://doi.org/10.1111%2Fj.1467-9280.2007.01917.x.

78 *If we think we're being shamed* Brandi Morris, Polymeros Chrysochou, Jacob Dalgaard Christensen, Jacob Orquin, Jorge Barraza, Paul Zak, and Panagiotis Mitkidis, "Stories vs. Facts: Triggering Emotion and Action-Taking on Climate Change," *Climatic Change* 154 (2019): 19–36, https://doi.org/10.1007/s10584-019-02425-6.

78 *"Puritan ethos of disapproval"* Rebecca Huntley, *How to Talk About Climate Change in a Way That Makes a Difference* (Sydney: Murdoch Books, 2020).

79 *anticipating the pride of making a choice* Claudia Schneider, Lisa Zaval, Elke Weber, and Ezra Markowitz, "The Influence of Anticipated Pride and Guilt on Pro-Environmental Decision Making," *PLOS One* 12, no. 11 (2017): e0188781, https://doi.org/10.1371/journal.pone.0188781.

79 *"Most women describe themselves"* https://potentialenergycoalition.org/.

79 *"I don't blame anyone"* Mary Annaïse Heglar, "I Work in the Environmental Movement. I Don't Care If You Recycle," *Vox*, June 4, 2019, https://www.vox.com/the-highlight/2019/5/28/18629833/climate-change-2019-green-new-deal.

79 *this system also provides our safety* Gabrielle Wong-Parodi and Irina Feygina, "Understanding and Countering the Motivated Roots of Climate Change Denial," *Current Opinion in Environmental Sustainability* 42 (2020): 60–65, https://doi.org/10.1016/j.cosust.2019.11.008.

82 *"Fear is not a great predictor"* As quoted in Rebecca Huntley, *How to Talk About Climate Change in a Way That Makes a Difference* (Sydney: Murdoch Books, 2020).

82 *"consistent with upholding"* Gabrielle Wong-Parodi and Irina Feygina, "Understanding and Countering the Motivated Roots of Climate Change Denial," *Current Opinion in Environmental Sustainability* 42 (2020): 60–65, https://doi.org/10.1016/j.cosust.2019.11.008.

82 *If our brain is hardwired* Tali Sharot, *The Influential Mind: What the Brain Reveals About Our Power to Change Others* (New York: Henry Holt and Co., 2017).

83 *Hannah Malcolm is a theologian* Hannah Malcolm, "Finding Words for the End of the World," *A Rocha Blog*, 2019, https://blog.arocha.org/en/finding-words-for-the-end-of-the-world/.

SECTION 3:
THE THREAT MULTIPLIER

CHAPTER 8

87 *"We are navigating recklessly"* Marcia Bjornerud, *Timefulness: How Thinking Like a Geologist Can Help Save the World* (Princeton: Princeton University Press, 2018).

88 *Arctic sea ice is declining* Florence Fetterer, Kenneth Knowles, Walter Meier, Matthew Savoie and Ann Windnagel, "Sea Ice Index, Version 3," *National Snow & Ice Data Center* (2016), https://doi.org/10.7265/N5736NV7.

88 *average thickness of the ice* Julienne Stroeve, Andrew Barrett, Mark Serreze, and Axel Schweiger, "Using records from submarine, aircraft and satellites to evaluate climate model simulations of Arctic sea ice thickness," *The Cryosphere* 8 (2014): 1839–1854, https://doi.org/10.5194/tc-8 -1839-2014.

89 *impact of warming temperatures on pests* David Lobell and Christopher Field, "Global Scale Climate–Crop Yield Relationships and the Impacts of Recent Warming," *Environmental Research Letters* 2, no. 1 (2007), https://iopscience.iop.org/article/10.1088/1748-9326/2/1/014002/meta.

90 *Bob Nash dismissed them* Texas Department of Public Safety Historical Museum and Research Center, *Twister!* (1970), https://texasarchive .org/2009_00879.

90 *one of the strongest tornadoes* Roger Edwards and Joe Shaefer, "Downtown Tornadoes," *Storm Prediction Center, NOAA/National Weather Service,* https://www.spc.noaa.gov/faq/tornado/downtown.html.

90 *hold up a snowball* Jeffrey Kluger, "Senator Throws Snowball! Climate Change Disproven!" *TIME,* February 27, 2015, https://time.com/3725994 /inhofe-snowball-climate/.

91 *Yale Program on Climate Communication* Jennifer Marlon, Peter Howe, Matto Mildenberger, Anthony Leiserowitz, and Xinran Wang, *Yale Climate Opinion Maps 2020,* https://climatecommunication.yale.edu/visu alizations-data/ycom-us/.

91 *the effects of psychological distance* Matthew Goldberg, Abel Gustafson, Seth Rosenthal, John Kotcher, Edward Maibach, and Anthony Leiserowitz, *For the First Time, the Alarmed Are Now the Largest of Global Warming's Six Americas* (New Haven: Yale Program on Climate Change Communication, 2020).

91 *"When it comes to rare probabilities"* Daniel Kahneman, *Thinking, Fast and Slow* (New York: Farrar, Straus and Giroux, 2013).

92 *they see it as distant and less relevant* Haoran (Chris) Chu and Janet

Yang, "Taking Climate Change Here and Now—Mitigating Ideological Polarization with Psychological Distance," *Global Environmental Change* 53 (2018): 174–181, https://doi.org/10.1016/j.gloenvcha.2018.09.013.

92 *local sea level rise affects us* Laurel Evans, Taciano L. Milfont, and Judy Lawrence, "Considering Local Adaptation Increases Willingness to Mitigate," *Global Environmental Change*, 25 (2014): 69–75, https://doi.org /10.1016/j.gloenvcha.2013.12.013.

95 *decline of civilizations* Harvey Weiss and Raymond Bradley, "What Drives Societal Collapse?" *Science* 291, no. 5504 (2001): 609–610, https:// doi.org/10.1126/science.1058775.

95 *over ten times faster* NASA Global Climate Change: Vital Signs of the Planet, "Climate Change: How Do We Know?" https://climate.nasa.gov /evidence/.

CHAPTER 9

97 *"Ours is the first generation"* Kate Raworth, *Doughnut Economics* (White River Junction: Chelsea Green Publishing, 2017).

97 *had a competition* Jennifer Loesch, "Voters declare Lubbock Toughest Weather City," *Lubbock Avalanche-Journal*, April 5, 2013, https://www .lubbockonline.com/article/20130405/NEWS/304059799.

97 *124 of these events* "NOAA National Centers for Environmental Information (NCEI) U.S. Billion-Dollar Weather and Climate Disasters (2020)," https://www.ncdc.noaa.gov/billions/, doi: 10.25921/stkw-7w73.

99 *four hundred all-time-high temperature records* Nassos Stylianou and Clara Guibourg, "Hundreds of Temperature Records Broken Over Summer," *BBC News*, October 9, 2019, https://www.bbc.com/news/science -environment-49753680.

100 *already risen by* Billy Sweet, Radley Horton, Robert Kopp, Allegra LeGrande, and Anastasia Romanou, "Sea level rise" in Climate Science Special Report: Fourth National Climate Assessment, Volume I, eds. Donald Wuebbles, David Fahey, Kathy Hibbard, David Dokken, Brooke Stewart, and Thomas Maycock (Washington, DC: U.S. Global Change Research Program, 2018), http://doi.org/10.7930/J01834ND.

100 *thirty percent faster* NASA, "Rising Waters: How NASA is Monitoring Sea Level Rise," (2020) https://www.nasa.gov/specials/sea-level-rise-2020/.

100 *levees were built to protect it* North Carolina State University, "Study Finds Flooding Damage to Levees Is Cumulative—and Often Invisible," *EurekAlert!*, January 2020, https://www.eurekalert.org/pub_releases/2020 -01/ncsu-sff012120.php.

101 *underwater in 1999* Richard M. Mizelle, Jr., "Princeville and the Environmen-

tal Landscape of Race," *Open Rivers*, spring 2016, https://editions.lib.umn.edu
/openrivers/article/princeville-and-the-environmental-landscape-of-race/.

101 *again in 2016* "Troubled Waters," Landslide 2018, Grounds for Democracy, https://tclf.org/sites/default/files/microsites/landslide2018/prince
ville.html.

000 *says resident Marvin Dancy* "Princeville fears being wiped out third
time," *WRAL*, September 10, 2018, https://www.wral.com/princeville
-fears-being-wiped-out-third-time/17833981/.

101 *700 million live* Barbara Neumann, Athanasios T. Vafeidis, Juliane Zimmermann, and Robert J. Nicholls, "Future Coastal Population Growth
and Exposure to Sea-Level Rise and Coastal Flooding—A Global Assessment," *PLOS One* 10, no. 6 (2015): e0131375. https://doi.org/10.1371
/journal.pone.0131375.

102 *oil and gas pipelines on thawing permafrost* Blake Sobczak, "Thawing
Permafrost Jeopardizes Massive Maze of Russian Pipelines," *E&E News*,
January 30, 2013, https://www.eenews.net/stories/1059975505.

102 *two hundred Native American villages in Alaska* U.S. Government Accountability Office (GAO), *Alaska Native Villages: Limited Progress Has
Been Made on Relocating Villages Threatened by Flooding and Erosion*,
GAO-09-551 (2019), https://www.gao.gov/products/GAO-09-551.

103 *cost of replacing these with all-season roads* Dan Levin, "Ice Roads Ease
Isolation in Canada's North, but They're Melting Too Fast," *New York
Times*, April 19, 2017, https://www.nytimes.com/2017/04/19/world/can
ada/ice-roads-ease-isolation-in-canadas-north-but-theyre-melting-too
-soon.html.

105 *most of the world's surface permafrost* Michael Meredith, Martin Sommerkorn, Sandra Cassotta, Chris Derksen, Aleksey Ekaykin, Anne Hollowed, Gary Kofinas, et al., "Polar Regions," in *IPCC Special Report on the
Ocean and Cryosphere in a Changing Climate* eds. Hans Otto Pörtner et al.
(Cambridge, UK and New York, NY: Cambridge University Press, 2019),
https://www.ipcc.ch/srocc/.

105 *the mental or existential distress* Glenn Albrecht, "Negating Solastalgia: An Emotional Revolution from the Anthropocene to the Symbiocene," *American Imago* 77, no. 1 (2020): 9–30, https://doi.org/10.1353
/aim.2020.0001.

107 *"we are all Kiribati"* Anote Tong and Matthieu Rytz, "Our Island Is
Disappearing but the President Refuses to Act," *Washington Post*, October 24, 2018, https://www.washingtonpost.com/news/theworldpost
/wp/2018/10/24/kiribati/.

CHAPTER 10

109 *"Every bit of warming matters"* "Intergovernmental Panel on Climate Change: Summary for Policymakers," in *Global Warming of 1.5°C.,* eds. Valerie Masson-Delmotte et al. (Cambridge: Cambridge University Press, 2018).

110 *summarized them in a study* Svante A. Arrhenius, "On the Influence of Carbonic Acid in the Air Upon the Temperature of the Ground," *Philosophical Magazine* 41 (1896): 237.

110 *over 420 ppm* Earth System Research Laboratories Global Monitoring Laboratory, "Carbon Cycle Greenhouse Gases: Trends in CO2," https://gml.noaa.gov/ccgg/trends/.

111 *Native American villagers in Alaska* U.S. Government Accountability Office (GAO), *Alaska Native Villages: Limited Progress Has Been Made on Relocating Villages Threatened by Flooding And Erosion* (2009), GAO-09-551. https://www.gao.gov/products/GAO-09-551.

111 *people who died in the heat wave of 2003* Jean-Marie Robine, Siu Lan K. Cheung, Sophie Le Roy, Herman Van Oyen, Clare Griffiths, Jean-Pierre Michel, and François Richard Herrmann, "Death Toll Exceeded 70,000 in Europe During the Summer of 2003," *Comptes Rendus Biologies* 33, no. 2 (2008): 171–178, https://doi.org/10.1016/j.crvi.2007.12.001.

111 *twice as likely as a result* Peter Stott, Dáithí Stone, and Myles Allen, "Human Contribution to the European Heatwave of 2003," *Nature* 432 (2004): 610–614, https://doi.org/10.1038/nature03089.

113 *"We're going to do some of each"* John Holdren, "Science and Technology for Sustainable Well-Being," *Science* 391, no. 5862 (2008): 424–434, https://doi.org/10.1126/science.1153386.

115 *The results of our work* Katharine Hayhoe, Daniel Cayan, Christopher B. Field, Peter C. Frumhoff, Edwin P. Maurer, Norman L. Miller, Susanne C. Moser, et al., "Emissions Pathways, Climate Change, and Impacts on California," *Proceedings of the National Academy of Sciences* 101, no. 34 (2004): 12422–12427, https://doi.org/10.1073/pnas.0404500101.

115 *executive order S-3-05* Governor Arnold Schwarzenegger, Executive Order S-3-05, June 1, 2005, https://static1.squarespace.com/static/549885d4e4b0ba0bff5dc695/t/54d7f1e0e4b0f0798cee3010/1423438304744/California+Executive+Order+S-3-05+(June+2005).pdf.

CHAPTER 11

117 *"Why do you fight for our planet?"* Gaurab Basu, Twitter post, November 30, 2020, 4:20 a.m., https://twitter.com/GaurabBasuMDMPH/status/1333370217125224450.

119 *"Americans tend to see global warming"* John Kotcher, Edward Maibach, Marybeth Montoro, and Susan Joy Hassol, "How Americans Respond to Information About Global Warming's Health Impacts: Evidence from a National Survey Experiment," *GeoHealth* 2, no. 9 (2018): 262–275, https://doi.org/10.1029/2018GH000154.

120 *the then hottest summer on record* Jean Robine, Jean-Marie, Siu Lan K. Cheung, Sophie Le Roy, Herman Van Oyen, Clare Griffiths, Jean-Pierre Michel, and François Richard Herrmann, "Death Toll Exceeded 70,000 in Europe During the Summer of 2003," *Comptes Rendus Biologies* 331, no. 2 (2008): 171–178, https://doi.org/10.1016/j.crvi.2007.12.001.

120 *"Hot temperatures increase aggression"* Craig Anderson, "Heat and Violence," *Current Directions in Psychological Research* 10, no. 1 (2001): 33–38, https://doi.org/10.1111%2F1467-8721.00109.

120 *"Failing to reduce heat-trapping emissions"* Kristina Dahl, Erika Spanger-Siegfried, Rachel Licker, Astrid Caldas, John Abatzoglou, Nicholas Mailloux, Rachel Cleetus, Shana Udvardy, Juan Declet-Barreto, and Pamela Worth, *Killer Heat in the United States: Climate Choices and the Future of Dangerously Hot Days* (Cambridge, MA: Union of Concerned Scientists, July 2019), https://www.ucsusa.org/sites/default/files/attach/2019/07/killer-heat-analysis-full-report.pdf.

120 *nearly 9 million premature deaths* Karn Vohra, Alina Vodonos, Joel Schwartz, Eloise Marais, Melissa Sulprizio, and Loretta Mickley, "Global mortality from outdoor fine particle pollution generated by fossil fuel combustion: Results from GEOS-Chem," *Environmental Research* 195 (2021), https://doi.org/10.1016/j.envres.2021.110754.

120 *"single largest environmental health risk"* Diarmid Campbell-Lendrum and Annette Prüss-Ustün, "Climate Change, Air Pollution and Non-communicable Diseases," *Bulletin of the World Health Organization* 97 (2019):160–161, http://dx.doi.org/10.2471/BLT.18.224295.

121 *even a 20 percent increase* Matt Cole, Ceren Ozgen, and Eric Strobl, "Air Pollution Exposure Linked to Higher COVID-19 Cases and Deaths—New Study," *The Conversation*, July 13, 2020, https://theconversation.com/air-pollution-exposure-linked-to-higher-covid-19-cases-and-deaths-new-study-141620.

121 *much more likely to die* Xiao Wu, Rachel Nethery, Benjamin Sabath, Danielle Braun, and Francesca Dominici, *Exposure to Air Pollution and COVID-19 Mortality in the United States* (Harvard University School of Public Health), https://projects.iq.harvard.edu/files/covid-pm/files/pm_and_covid_mortality.pdf.

121 *Harvard researchers believe* Alexandra Sternlicht, "Higher Coronavirus Mortality Rates for Black Americans and People Exposed to Air Pollution,"

Forbes, April 7, 2020, https://www.forbes.com/sites/alexandrasternlicht /2020/04/07/higher-coronavirus-mortality-rates-for-people-exposed-to -air-pollution-black-americans/#35af0ç5a362f.

121 *increase the risk of dementia* Ruth Peters, Nicole Ee, Jean Peters, Andrew Booth, Ian Mudway and Kaarin Anstey, "Air Pollution and Dementia: A Systematic Review," *Journal of Alzheimers Disease* 70 (2019): S145–S163, https://doi.org/10.3233/JAD-180631.

121 *affect the newly developing brains* Healthy Babies Bright Futures and George Mason University Center for Climate Change Communication, "The Link Between Fossil Fuels and Neurological Harm," Healthy Babies Bright Future (HBBF), 2019, https://www.hbbf.org/sites/default/files/docu ments/2018-06/LinkFossilFuelsNeurologicaHarm_04-06-18_v2.pdf.

121 *delaying or impairing cognitive development* John Kotcher, Edward Maibach, and Wen-Tsing Choi, "Fossil Fuels Are Harming Our Brains: Identifying Key Messages About the Health Effects of Air Pollution from Fossil Fuels," *BMC Public Health* 19 (2019): 1079, https://doi.org/10.1186 /s12889-019-7373-1.

121 *an ambitious Climate Action Plan for Chicago* City of Chicago, "Chicago Climate Action Plan," Chicago.gov, 2020, https://www.chicago.gov/city /en/progs/env/climateaction.html.

122 *they identified "hot spots"* Climate Change Adaptation Resource Center (ARC-X), "Chicago, IL Uses Green Infrastructure to Reduce Extreme Heat," EPA.gov, https://www.epa.gov/arc-x/chicago-il-uses-green-infra structure-reduce-extreme-heat.

122 *one of the largest electric bus fleets* Monica Eng, Jessica Pupovac, and Mackenzie Crosson, "How Is Chicago Doing On Its Ambitious 2020 Climate Goals?," WBEZ 91.5 Chicago, June 2, 2019, http://interactive.wbez .org/curiouscity/climate-goals/?_ga=2.160216500.887932196.15588811 95-255549137.1543244588.

122 *knew or should have known* Christina Davis, "Farmers Insurance Class Action Lawsuit Places Home Flooding Blame on City of Chicago," Top Class Actions, April 30, 2014, https://topclassactions.com/lawsuit-set tlements/lawsuit-news/25162-farmers-insurance-class-action-lawsuit -places-home-flooding-blame-city-chicago/.

122 *two new reservoirs* Craig Dellimore, "Massive New Reservoir to Help Alleviate Chicago Area Flooding," CBS Chicago, September 1, 2015, https:// chicago.cbslocal.com/2015/09/01/massive-new-reservoir-to-help-allevi ate-chicago-area-flooding/.

123 *3 million people across the globe* World Health Organization, *WHO World Water Day Report*, https://www.who.int/water_sanitation_health /takingcharge.html.

123 *nearly 40 percent more rainfall* Mark Risser and Michael Wehner, "Attributable Human-Induced Changes in the Likelihood and Magnitude of the Observed Extreme Precipitation During Hurricane Harvey," *Geophysical Research Letters* 44, no. 24 (2017): 12,457–12,464, https://doi.org/10.1002/2017GL075888.

123 *four times the economic damage* David Frame, Michael Wehner, Ilan Noy, and Suzanne Rosier, "The economic costs of Hurricane Harvey attributable to climate change," *Climatic Change* 160 (2020): 271–281, https://doi.org/10.1007/s10584-020-02692-8.

123 *skin infections to diarrheal disease* Juanita Constible, "The Emerging Public Health Consequences of Hurricane Harvey," NRDC.org, August 29, 2018, https://www.nrdc.org/experts/juanita-constible/emerging-public-health-consequences-hurricane-harvey.

123 *a significant jump in cholera cases* Erin Hulland, Saleena Subaiya, Katilla Pierre, Nickolson Barthelemy, Jean Samuel Pierre, Amber Dismer, Stanley Juin, David Fitter, and Joan Brunkard, "Increase in Reported Cholera Cases in Haiti Following Hurricane Matthew: An Interrupted Time Series Model," *American Journal of Tropical Medicine and Hygiene* 100, no. 2 (2019): 368–373, https://doi.org/10.4269/ajtmh.17-0964.

124 *access to safe sanitation* United Nations, "Goal 6: Ensure Availability and Sustainable Management of Water and Sanitation for All," sdgs.un.org, https://sdgs.un.org/goals/goal6.

124 *"Water can help fight climate change"* United Nations, "World Water Day Focusing on the Importance of Water and Climate Change," *Sustainable Development Goals*, March 22, 2020, https://sustainabledevelopment.un.org/index.php?page=view&type=13&nr=3280&menu=1634.

124 *Sanergy is an organization* Sanergy, "Bold Solutions for Booming Cities: The Urban Sanitation Challenge," Sanergy.com, http://www.sanergy.com/.

124 *Sulabh International* Sulabh International Social Service Organisation, "Our Work," sulabhinternational.org, https://www.sulabhinternational.org/sulabh-technologies-bio-gas/.

124 *turning human waste into biogas* Melanie Sevcenko, "Power to the Poop: One Colorado City Is Using Human Waste to Run Its Vehicles," *Guardian*, January 16, 2016, https://www.theguardian.com/environment/2016/jan/16/colorado-grand-junction-persigo-wastewater-treatment-plant-human-waste-renewable-energy.

124 *In 2020 the Puerto Rico* Allen Brown, "Energy Insurrection," *The Intercept*, February 9, 2020, https://theintercept.com/2020/02/09/puerto-rico-energy-electricity-solar-natural-gas/.

125 *"climate, environmental degradation and natural disasters"* United Na-

tions, *Global Compact on Refugees* (New York: United Nations, 2018), https://www.unhcr.org/the-global-compact-on-refugees.html.

125 *displacing some 20 million people* Oxfam International, "Forced from Home: Climate-Fuelled Displacement," Oxfam, December 2, 2019, https://www.oxfam.org/en/research/forced-home-climate-fuelled-dis placement?cid=aff_affwd_donate_id201309&awc=5991_1580878311_8 6c6eba9dd6c9035f43d0e8c338aecb9.

125 *between twelve and thirty-five thousand tons* S. Saatchi, R. Houghton, R. Dos Santos Alvala, J. Soares, and Y. Yu, "Distribution of aboveground live biomass in the Amazon basin," *Global Change Biology* 13, no. 4 (2007): 816–837, https://doi.org/10.1111/j.1365-2486.2007.01323.x.

126 *reduced deforestation by 30 percent* Paul Ferraro and Rhita Simorangkir, "Conditional Cash Transfers to Alleviate Poverty Also Reduced Deforestation in Indonesia," *Science Advances* 6, no. 24 (2020): eaaz1298, https://doi.org/10.1126/sciadv.aaz1298.

126 *programs in Nepal, China, and India* "Poverty Reduction and Forest Protection," The Borgen Project, https://borgenproject.org/poverty-reduc tion-and-forest-protection/.

126 *Oxford Dictionary defines* Oxford Languages, "Word of the Year 2019," *Oxford University Press*, https://languages.oup.com/word-of-the-year/2019/.

126 *jumped a stunning 4,000 percent* Ibid.

127 *The 2020 Lancet Countdown* The *Lancet* Countdown 2020 Report, LancetCountdown.org, https://www.lancetcountdown.org/2020-report/.

127 *50 percent of people surveyed in the U.S.* John Kotcher, Edward Maibach, Seth Rosenthal, Abel Gustafson, and Anthony Leiserowitz, *Americans increasingly understand that climate change harms human health.* (New Haven: Yale Program on Climate Change Communication, 2020).

128 *"Preservation of our environment"* Ronald Reagan, Remarks on Signing the Annual Report of the Council on Environmental Quality, July 11, 1984, https://www.reaganlibrary.gov/research/speeches/71184a.

128 *As Christiana Figueres says* Christiana Figueres and Tom Rivett-Carnac, *The Future We Choose: Surviving the Climate Crisis* (New York: Alfred A. Knopf, 2020).

SECTION 4:
WE CAN FIX IT

CHAPTER 12

131 *"Denialism . . . "* Alastair McIntosh, *Riders on the Storm: The Climate Crisis and the Survival of Being* (Edinburgh: Birlinn, 2020).

132 *I scored them* Katharine Hayhoe and Andrew Leach, "How The Four Federal Parties' Climate Plans Stack Up," *Chatelaine*, October 3, 2019, https://www.chatelaine.com/living/politics/2019-federal-election-climate/.

133 *more than 10 percent* Mark Kane, "10% of Norway's Passenger Vehicles are Plug Ins," InsideEVs, November 7, 2018, https://insideevs.com/news/341060/10-of-norways-passenger-vehicles-are-plug-ins/.

133 *were zero-emission vehicles* Charles Riley, "Electric cars hit record 54% of sales in Norway as VW overtakes Tesla," *CNN Business*, January 5, 2021, https://www.cnn.com/2021/01/05/business/norway-electric-cars-vw-tesla/index.html.

133 *voted to divest* Dieter Holger, "Norway's Sovereign-Wealth Fund Boosts Renewable Energy, Divests Fossil Fuels," *The Wall Street Journal*, June 12, 2019, https://www.wsj.com/articles/norways-sovereign-wealth-fund-boosts-renewable-energy-divests-fossil-fuels-11560357485

133 *Kari interviewed dozens of Norwegians* Kari Norgaard, *Living in Denial: Climate Change, Emotions, and Everyday Life* (Cambridge, MA: The MIT Press, 2011).

134 *This counterintuitive term was first applied* Troy Campbell and Aaron Kay, "Solution Aversion: On the Relation Between Ideology and Motivated Disbelief," *Journal of Personality and Social Psychology* 107, no. 5 (2014): 809–824, https://doi.org/10.1037/a0037963.

136 *According to the* Carbon Majors Report Climate Accountability Institute, the Carbon Majors Database, 2017, https://b8f65cb373b1b7b15feb-c70d8ead6ced550b4d987d7c03fcdd1d.ssl.cf3.rackcdn.com/cms/reports/documents/000/002/327/original/Carbon-Majors-Report-2017.pdf?1499691240.

137 *"Exxon knew about climate change"* https://exxonknew.org/.

137 *"The CO$_2$ concentration in the atmosphere"* Geoffrey Supran and Naomi Oreskes, "Assessing ExxonMobil's Climate Change Communications (1977–2014)," *Environmental Research Letters* 12, no. 8 (2017), https://doi.org/10.1088/1748-9326/.

137 *her eye-opening 2010 book* Naomi Oreskes and Eric Conway, *Merchants of Doubt: How a Handful of Scientists Obscured the Truth on Issues from Tobacco Smoke to Global Warming* (New York: Bloomsbury Publishing, 2011).

137 *Jim Hoggan skillfully dissects in his 2009 book* James Hoggan, *Climate Cover-up: The Crusade to Deny Global Warming* (Vancouver, Canada: Greystone Books, 2009).

138 *Full-page ads in prominent newspapers* Hiroko Tabuchi, "How One Firm Drove Influence Campaigns Nationwide for Big Oil," *New York Times*, November 11, 2020, https://www.nytimes.com/2020/11/11/climate/fti-consulting.html.

138 *donations to politicians at every level* Public Accountibility Institute, *The Money Behind Empower Texans*, September 25, 2019, https://public-accountability.org/report/the-money-behind-empower-texans/.

138 *By April 2019* A. Gustafson, S. Rosenthal, P. Bergquist, M. Ballew, M. Goldberg, J. Kotcher, A. Leiserowitz, and E. Maibach, *Changes in Awareness of and Support for the Green New Deal: December 2018 to April 2019* (New Haven, CT: Yale Program on Climate Change Communication, 2019), doi: 10.17605/OSF.IO/P8ZBN.

140 *"connected to a system of production"* Noel Healy, "Blood Coal: Ireland's Dirty Secret," *Guardian*, October 25, 2018, https://www.theguardian.com/environment/climate-consensus-97-per-cent/2018/oct/25/blood-coal-irelands-dirty-secret.

140 *"Oftentimes, residents . . . live in agonizing conditions"* Emem Edoho, "Oil Spills in Nigeria: Health Risks and Environmental Degradation," https://www.gndr.org/programmes/advocacy/365-disasters/more-than-365-disasters-blogs/item/1450-oil-spills-nigeria.html.

141 *deaths are in the U.S.* Fabio Caiazzo, Akshay Ashok, Ian, A. Waitz, Steve H. L. Yim, and Steven Barrett, "Air Pollution and Early Deaths in the United States, Part I: Quantifying the Impact of Major Sectors in 2005," *Atmospheric Environment* 79 (2013): 198-208, https://doi.org/10.1016/j.atmosenv.2013.05.081.

141 *climate change has already increased the economic gap* Marshall Burke and Noah Diffenbaugh, "Global Warming Has Increased Global Economic Inequality," *Proceedings of the National Academy of Sciences* 116 no. 20 (2019): 9808–9813, https://doi.org/10.1073/pnas.1816020116.

141 *negated over fifty years of advances* United Nations Human Rights Council, "U.N. Expert Condemns Failure to Address Impact of Climate Change on Poverty," June 25, 2019, https://www.ohchr.org/EN/NewsEvents/Pages/DisplayNews.aspx?NewsID=24735&LangID=E.

141 *a 75 percent drop in average income* Marshall Burke, Solomon Hsiang, and Edward Miguel, "Global Non-linear Effect of Temperature on Economic Production," *Nature* 527 (2015): 235–239, https://doi.org/10.1038/nature15725.

141 *In Kenya, for example, women farmers* World Bank, Kenya Agricultural Productivity and Agribusiness Project, 2015, https://projects.worldbank.org/en/projects-operations/project-detail/P109683?lang=en.

141 *In Mali, women who attend* UNESCO, "Education Counts: Towards the Millennium Development Goals," 2010, https://unesdoc.unesco.org/ark:/48223/pf0000190214_eng.

142 *For each additional year of schooling* John Cleland, "Survival in Developing Countries," *Social Science & Medicine* 27, no. 12 (1988): 1357–1368.

142 *10 percent in Malawi* Liliana Adriano and Christiaan Monden, "The Causal Effect of Maternal Education on Child Mortality: Evidence from a Quasi-Experiment in Malawi and Uganda," *Demography* 56 (2019): 1765–1790.

142 *50 percent more likely to survive* UNESCO, "Education Counts: Towards the Millennium Development Goals," 2010, https://unesdoc.unesco.org/ark:/48223/pf0000190214_eng.

142 *"By this everyone will know"* John 13:15 (New International Version).

142 *"the only thing that counts"* Galatians 5:6 (New International Version).

CHAPTER 13

143 *"As we retreat"* Eric Liu, *You're More Powerful Than You Think: A Citizen's Guide to Making Change Happen* (New York: Public Affairs/Perseus Books, 2017).

144 *The fundamental concept of a "commons"* William Forster Lloyd, *Two Lectures on the Checks to Population* (Oxford: Oxford University, 1833).

144 *"tragedy of the commons"* Garrett Hardin, "The Tragedy of the Commons," *Science* 162, no. 3859 (1968): 1243–1248.

144 *Elinor Ostrom received the Nobel Memorial Prize* The Nobel Prize, "Elinor Ostrom: Facts," 2009, https://www.nobelprize.org/prizes/economic-sciences/2009/ostrom/facts/.

144 *expressed eugenic views* Matto Mildenberger, "The Tragedy of the Tragedy of the Commons," *Scientific American*, April 23, 2019, https://blogs.scientificamerican.com/voices/the-tragedy-of-the-tragedy-of-the-commons/.

145 *so many do in the U.S.* Edelman, "Edelman 2020 Trust Barometer," January 19, 2020, https://www.pewresearch.org/politics/2019/04/11/public-trust-in-government-1958-2019/.

145 *recent polls indicate* Pew Research Center, "Public Trust in Government: 1958–2019," April 11, 2019, https://www.pewresearch.org/politics/2019/04/11/public-trust-in-government-1958-2019/.

146 *a prescient report* Worldwide Fund for Nature, "The Loss of Nature and the Rise of Pandemics: Protecting Human and Planetary Health," WWF International, March 2020, https://wwfeu.awsassets.panda.org/downloads/the_loss_of_nature_and_rise_of_pandemics___protecting_human_and_planetary_health.pdf.

146 *one in six premature deaths* Philip Landrigan, Richard Fuller, Nereus J. R. Acosta, Olusoji Adeyi, Robert Arnold, Niladri Basu, et al., "The Lancet Commission on Pollution and Health," *Lancet* 391, no. 10119 (October 19, 2017): https://doi.org/10.1016/S0140-6736(17)32345-0.

147 *Betsy Hartmann points out* Betsy Hartmann, *Reproductive Rights and Wrongs: The Global Politics of Population Control* (Chicago: Haymarket Books, 2016).

147 *Global Footprint Network's calculator* Ecological Footprint Explorer, https://data.footprintnetwork.org/#/, accessed September 2020.

148 *16 metric tons of carbon dioxide* U.S. Environmental Protection Agency, *Inventory of U.S. Greenhouse Gas Emissions and Sinks*, https://www.epa.gov/ghgemissions/inventory-us-greenhouse-gas-emissions-and-sinks.

148 *Australians produce* Union of Concerned Scientists, "Each Country's Share of CO_2 Emissions," https://www.ucsusa.org/resources/each-countrys-share-co2-emissions.

148 *driving a midsized car* U.S. Environmental Protection Agency, "Greenhouse Gas Emissions from a Typical Passenger Vehicle," https://www.epa.gov/greenvehicles/greenhouse-gas-emissions-typical-passenger-vehicle.

148 *the richest 10 percent of people* Tim Gore, *Confronting Carbon Inequality*, Oxfam International, September 21, 2020, https://www.oxfam.org/en/research/confronting-carbon-inequality.

148 *It's the U.S. military* Neta Crawford, *Pentagon Fuel Use, Climate Change, and the Costs of War*, Watson Institute (Providence, RI: Brown University, 2019), https://watson.brown.edu/costsofwar/files/cow/imce/papers/Pentagon%20Fuel%20Use%2C%20Climate%20Change%20and%20the%20Costs%20of%20War%20Revised%20November%202019%20Crawford.pdf.

149 *extracted and popularized* Mark Kaufman, "The carbon footprint sham," *Mashable*, July 17, 2020, https://mashable.com/feature/carbon-footprint-pr-campaign-sham/.

149 *announce a carbon-neutral goal* British Petroleum, "BP sets amibition for net zero by 2050, fundamentally changing organisation to deliver," February 12, 2020, https://www.bp.com/en/global/corporate/news-and-insights/press-releases/bernard-looney-announces-new-ambition-for-bp.html.

150 *told a group of CEOs* Emily Gosden, "Strawberries in winter takes the biscuit, says Shell boss," *The Times UK*, June 12, 2019, https://www.thetimes.co.uk/article/strawberries-in-winter-takes-the-biscuit-says-shell-boss-t9c7w79dk.

150 *over 8,000 million tons* Richard Heede, "Tracing anthropogenic carbon dioxide and methane emissions to fossil fuel and cement producers, 1854–2010," *Climatic Change*, 122 (2014): 229–241, https://doi.org/10.1007/s10584-013-0986-y.

150 *200 billion trees* Calculation based on Joseph Veldman, Julie Aleman, Swanni Alvarado, Michael Anderson, Sally Archibald et al., "Comment on 'The global tree restoration potential,'" *Science*, 366, no. 6463 (2019), https://doi.org/10.1126/science.aay7976.

150 *more than five times* J. Elliott Campbell, David Lobell, Robert Genova, and Christopher Field, "The Global Potential of Bioenergy on Abandoned Agricultural Lands," *Environmental Science & Technology*, 42, no. 15 (2008): 5791–5794, https://doi.org/10.1021/es800052w.

CHAPTER 14

151 *"Walking out is not"* Mary Annaïse Heglar, "Home is Always Worth It" in *All We Can Save: Truth, Courage, and Solutions for the Climate Crisis* (New York: One World/Random House, 2020).

152 *India planned to* Republic of India, Ministry of Environment, Forest, and Climate Change, *India's Intended Nationally Determined Contribution: Working Towards Climate Justice*, https://www4.unfccc.int/sites/ndcstaging/PublishedDocuments/India%20First/INDIA%20INDC%20TO%20UNFCCC.pdf.

152 *had already capped* Nina Chestney and Pete Harrison, "EU carbon cap greater under 30 pct emissions cut," *Reuters*, April 30, 2010, https://www.reuters.com/article/us-emissions-trading-europe-idUKTRE63T3H820100430.

152 *Bhutan was preserving* Royal Government of Bhutan, Ministry of Agriculture and Forests, "Press Release on COP21," December 10, 2015, http://www.moaf.gov.bt/press-release-on-cop21/.

152 *the reductions we need* "Climate Action Tracker," https://climateaction-tracker.org/, accessed September 2020.

152 *bringing nothing to the table* Ibid.

152 *provide economic incentives* Niven Winchester, "Can Tariffs Be Used to Enforce Paris Climate Commitments?," *World Economy* 41, no. 10 (2018): 2650–2668, https://doi.org/10.1111/twec.12679.

154 *the impacts of 1.5 versus 2°C* Intergovernmental Panel on Climate Change. Summary for Policymakers. In: *Global Warming of 1.5°C*, eds. Valerie Masson-Delmotte et al. (Cambridge: Cambridge University Press, 2018).

154 *"The climate has changed and is always changing"* Stephen Leahy, "Climate Science Report Contradicts Trump Administration Positions," *National Geographic*, November 4, 2017. https://www.nationalgeographic.com/news/2017/11/climate-change-usa-government-science-environment/.

154 *"largely based on the most extreme scenario"* James Cook, "Sounding an Alarm," *BBC News*, November 24, 2018, https://www.bbc.com/news/world-us-canada-46325168.

155 *"The observed increase in global carbon"* Katharine Hayhoe, James Edmonds, Robert Kopp, Allegra LeGrande, Benjamin Sanderson, Michael Wehner, and Donald Wuebbles, "Climate models, scenarios, and pro-

jections," in *Climate Science Special Report: Fourth National Climate Assessment, Volume I*, eds. Donald Wuebbles, David Fahey, Kathy Hibbard, David Dokken, Brooke Stewart, and Thomas Maycock (Washington, DC: U.S. Global Change Research Program, 2018), http://doi.org/10.7930 /J0WH2N54.

155 *the We Are Still In movement* https://www.wearestillin.com/, accessed September 2020.

155 *the Houston Climate Action Plan* City of Houston Climate Action Plan, http://www.greenhoustontx.gov/climateactionplan/index.html, accessed September 2020.

156 *the European Union set up* European Commission, "EU Emissions Trading System," https://ec.europa.eu/clima/policies/ets_en.

156 *were estimated to be 21 percent lower than in 2005* Ibid.

156 *a cap-and-trade approach successfully helped* Gabriel Chan, Robert N. Stavins, Robert C. Stowe, and Richard Sweeney, *The SO2 Allowance Trading System and the Clean Air Act Amendments of 1990: Reflections on Twenty Years of Policy Innovation* (Cambridge, MA: Harvard Environmental Economics Program, 2012).

156 *reduce carbon emissions in the U.S. Northeast* The Regional Greenhouse Gas Initiative, https://www.rggi.org.

000 *California since 2013* California Air Resources Board, "Cap-and-Trade Program," https://ww2.arb.ca.gov/our-work/programs/cap-and-trade-program.

156 *economist Kate Raworth explains* Kate Raworth, *Doughnut Economics* (White River Junction: Chelsea Green Publishing, 2017).

157 *the U.S. already effectively operates* The World Bank, "Carbon Pricing Dashboard," https://carbonpricingdashboard.worldbank.org/map _data.

157 *a carbon price of $17* Kepos Carbon Barometer, https://www.carbonbarometer.com/#/#carbonBarometer.

157 *carbon is currently priced* Energy Hub, "Canadian Carbon Pricing Mechanisms," https://www.energyhub.org/carbon-pricing/.

157 *Globally, the average price* Ian Parry, *Putting a Price on Pollution*, International Monetary Fund, December 2019, https://www.imf.org/external /pubs/ft/fandd/2019/12/pdf/the-case-for-carbon-taxation-and-putting -a-price-on-pollution-parry.pdf.

157 *Nordhaus's model* William Nordhaus, "An Optimal Transition Path for Controlling Greenhouse Gases," *Science* 258, no. 5086 (1992): 1315–1319, https://doi.org/10.1126/science.258.5086.1315.

158 *decreased by about 0.25 percent* Rohan Best, Paul Burke, and Frank Jotzo, "Carbon Pricing Efficacy: Cross-Country Evidence," *Environmental and*

Resource Economics 77 (2020): 69–94, https://doi.org/10.1007/s10640-020 -00436-x.

158 **British Columbia introduced a price on carbon in 2008** "British Columbia's Carbon Tax," https://www2.gov.bc.ca/gov/content/environment/climate-change/planning-and-action/carbon-tax, accessed September 2020.

158 **decreased by more than 17 percent** Leyland Cecco, "How to Make a Carbon Tax Popular? Give the Proceeds to the People," *Guardian*, December 4, 2018, https://www.theguardian.com/world/2018/dec/04/how-to-make -a-carbon-tax-popular-give-the-profits-to-the-people.

158 **the decrease came from efficiency improvements** Dave Sawyer and Seton Stiebert, "The True Measure of BC's Carbon Tax," *Policy Options*, May 2, 2019, https://policyoptions.irpp.org/magazines/may-2019/true-measure -bcs-carbon-tax/.

159 **Personal provincial income tax rates dropped** Stewart Elgie, "British Columbia's Carbon Tax Shift: An Environmental and Economic Success," *World Bank Blogs*, September 10, 2014, https://blogs.worldbank.org/climatechange/british-columbia-s-carbon-tax-shift-environmental-and -economic-success.

159 **60 percent in total** Rachel Maclean, "Alberta's Carbon Tax Brought in Billions. See Where It Went," CBC, April 8, 2019, https://www.cbc.ca/news /canada/calgary/carbon-tax-alberta-election-climate-leadership-plan -revenue-generated-1.5050438.

159 **four provinces with a price on carbon** Julia-Maria Becker and Maximilian Kniewasser, "These Provinces Led in Economic Growth. They Also Price Carbon Pollution," *Hill Times*, January 17, 2018, https://www.hilltimes .com/2018/01/17/provinces-led-economic-growth-also-price-carbon -pollution/131297.

159 **22 percent of global greenhouse gas emissions** World Bank, "Carbon Pricing Dashboard," https://carbonpricingdashboard.worldbank.org/, accessed September 2020.

159 **"an international policy institute"** Climate Leadership Council, https:// clcouncil.org/, accessed September 2020.

159 **to cut carbon emissions by 57 percent by 2035** Catrina Rorke and Greg Bertelsen, "America's Climate Advantage," Climate Leadership Council, September 2020, https://clcouncil.org/reports/americas-carbon-advantage.pdf.

160 **"the green paradox"** Hans-Werner Sinn, *The Green Paradox: A Supply-Side Approach to Global Warming* (Cambridge, MA: MIT Press, 2012).

160 **China's coal consumption largely plateaued** British Petroleum, "Statistical Review of World Energy 2020," https://www.bp.com/content/dam/bp /business-sites/en/global/corporate/pdfs/energy-economics/statistical -review/bp-stats-review-2020-full-report.pdf.

160 *building hundreds of coal-fired power plants* Steve Inskeep and Ashley Westerman, "Why Is China Placing a Global Bet on Coal?," *NPR World*, April 29, 2019, https://www.npr.org/2019/04/29/716347646/why-is-china -placing-a-global-bet-on-coal.

160 *countries, like Pakistan and Vietnam* Richard Talley, "Why Is China Funding Unsustainable Coal Projects in Pakistan?," OilPrice.com, April 27, 2017, https://oilprice.com/Energy/Energy-General/Why-Is-China-Funding -Unsustainable-Coal-Projects-In-Pakistan.html.

CHAPTER 15

161 *"Energy managed wisely"* Michael Webber, *Power Trip: The Story of Energy* (New York: Basic Books/Perseus Books, 2019).

161 *Energy poverty is real* International Energy Agency, SDG7: Data and Projections (Paris: IEA, 2020) https://www.iea.org/reports/sdg7-data-and -projections.

161 *Gladys lives in a small community* Solar Sister, "Impact Story: Gladys," https://solarsister.org/impact-story/gladys/, accessed September 2020.

163 *The Nations own 51 percent* Energy and Mines, "AurCrest Gold Signs Letter of Intent with Cat Lake First Nation to Develop up to 40 MWs of Renewable Energy," March 23, 2016, https://energyandmines.com/2016/03 /aurcrest-gold-signs-letter-of-intent-with-cat-lake-first-nation-to-de velop-up-to-40-mws-of-renewable-energy/.

163 *the opportunity to benefit from the mine* Northern Ontario Business, "Indigenous Gold Explorer Looks to Power Up First Nation Communities," September 25, 2018, https://www.northernontariobusiness.com /industry-news/aboriginal-businesses/indigenous-gold-explorer-looks -to-power-up-first-nation-communities-1061456.

163 *don't have abundant reserves* British Petroleum, "Statistical Review of World Energy 2019," https://www.bp.com/en/global/corporate/energy -economics/statistical-review-of-world-energy.html.

164 *more electricity globally than coal* International Energy Agency, https:// iea.blob.core.windows.net/assets/3350006e-c203-4b21-aa45-faefd36f 22ad/Renewables2020-ExecutiveSummary.pdf.

164 *the top five emerging markets* BloombergNEF, "ClimateScope 2019," https://global-climatescope.org/.

164 *these countries are leading the charge* BloombergNEF, "Developing Nations Assume Mantle of Global Clean Energy Leadership," November 27, 2018, https://about.bnef.com/blog/developing-nations-assume-mantle -global-clean-energy-leadership/.

164 *70 percent of new electricity* International Renewable Energy Agency,

"Renewable Capacity Statistics 2020," March 2020, https://www.irena.org
/publications/2020/Mar/Renewable-Capacity-Statistics-2020.

164 *soared to over 90 percent* International Energy Agency, *Renewables 2020*,
November 2020, https://www.iea.org/reports/renewables-2020.

164 *"climate change is an opportunity"* Richard Schiffmann. "How Can We
Make People Care About Climate Change?," *Yale360*, July 9, 2015, https://
e360.yale.edu/features/how_can_we_make_people_care_about_climate
_change.

165 *Almost 23 percent of the electricity* Electric Reliability Council of
Texas, "Fact Sheet February 2021," http://www.ercot.com/content/wcm
/lists/219736/ERCOT_Fact_Sheet_2.12.21.pdf.

165 *countries at or near 100 percent clean energy* International Renewable Energy Agency, "Renewable Capacity Statistics 2020," March 2020, https://www
.irena.org/publications/2020/Mar/Renewable-Capacity-Statistics-2020.

165 *prices for renewable energy* Mark Chediak and Brian Eckhouse, "Solar
and Wind Power So Cheap They're Outgrowing Subsidies," *Bloomberg*, September 19, 2019, https://www.bloomberg.com/news/features
/2019-09-19/solar-and-wind-power-so-cheap-they-re-outgrowing-sub
sidies?srnd=premium.

165 *by 2019 they made up 18 percent* United Nations Environment Program,
"Global Trends in Renewable Energy Investment 2019," September 11,
2019, https://www.unenvironment.org/resources/report/global-trends
-renewable-energy-investment-2019.

165 *fossil fuel use is subsidized* David Coady, Ian Parry, Nghia-Piotr Le,
and Baoping Shang, *Global Fossil Fuel Subsidies Remain Large: An Update Based on Country-Level Estimates*, International Monetary Fund
working paper, May 2, 2019, https://www.imf.org/en/Publications/WP
/Issues/2019/05/02/Global-Fossil-Fuel-Subsidies-Remain-Large-An-Up
date-Based-on-Country-Level-Estimates-46509.

166 *slightly more than the Pentagon's budget* National Priorities Project, "Federal Budget 101," https://www.nationalpriorities.org/campaigns/military
-spending-united-states/, accessed September 2020.

166 *the entire U.S. power grid* Amol Phadke, Umed Paliwal, Nikit Abhyankar,
Taylor McNair, Ben Paulos, David Wooley, and Ric O'Connell, *2035 The
Report* (Goldman School of Public Policy, University of California Berke-
ley, 2020), https://www.2035report.com/.

166 *"for every $1 spent for the energy transition"* International Renewable En-
ergy Agency, "Renewable Capacity Statistics 2019," March 2019, https://
www.irena.org/publications/2019/Mar/Renewable-Capacity-Statistics-2019.

166 *the U.S. military has spent more than $5 trillion* Mark Thompson, "Add-
ing Up the Cost of Our Never-Ending Wars," POGO (Project on Gov-

ernment Oversight), December 17, 2019, https://www.pogo.org/analysis /2019/12/adding-up-the-cost-of-our-never-ending-wars/.

166 *Method Soap's factory outside Chicago* https://methodhome.com/be yond-the-bottle/soap-factory/, accessed September 2020.

166 *bought more renewable energy than anyone else* Jillian Ambrose, "Tech Giants Power Record Surge in Renewable Energy Sales," *Guardian*, January 28, 2020, https://www.theguardian.com/environment/2020/jan/28 /google-tech-giants-spark-record-rise-in-sales-of-renewable-energy.

166 *committed to achieving 100 percent clean energy* https://www.there100 .org/re100-members, accessed November 2020.

167 *Canada and the U.S. are tied at number ten* American Council for an Energy-Efficient Economy, "The International Energy Efficiency Scorecard 2018," https://www.aceee.org/portal/national-policy/international -scorecard.

167 *cut U.S. carbon emissions in half by 2050* Lowell Unger and Steven Nadel, "Halfway There: Energy Efficiency Can Cut Energy Use and Greenhouse Gas Emissions in Half by 2050," American Council for an Energy-Efficient Economy, September 18, 2019, https://www.aceee.org/research-report/u 1907.

167 *physicist Saul Griffith has a plan* Saul Griffith, Sam Calisch, and Laura Fraser, *Rewiring America, a Handbook for Winning the Climate Fight* (Rewiring America: 2020), www.rewiringamerica.org/handbook.

167 *the biggest chunk of global transportation emissions* IEA, *Energy Technology Perspectives 2020*, September 2020, https://www.iea.org/reports /energy-technology-perspectives-2020.

167 *Volvo announced recently* "Volvo Cars Reveals Ambitious New Climate Plan," November 22, 2019, https://www.volvocars.com/au/About /Australia/I-Roll-eNewsletter/2019/November/Volvo-Cars-reveals-ambi tious-new-climate-plan.

168 *General Motors proclaimed* Sam Abuelsamid, "GM To Make Only Electric Vehicles by 2035, Be Carbon Neutral By 2040," *Forbes*, January 28, 2021, https://www.forbes.com/sites/samabuelsamid/2021/01/28/gen-eral-motors-commits-to-being-carbon-neutral-by-2040/?sh=44cf5d 656355.

168 *As of 2020, twenty countries* Kevin Joshua Ng, "List of Countries Banning Internal Combustion Engines in the Near Future," *eCompareMo.com*, https://www.ecomparemo.com/info/list-of-countries-banning-internal -combustion-engines-in-the-near-future.

168 *Industry is responsible* U.S. Environmental Protection Agency, "Global Greenhouse Gas Emissions Data," https://www.epa.gov/ghgemissions /global-greenhouse-gas-emissions-data.

168 *a new start-up called Heliogen* https://heliogen.com/, accessed September 2020.

169 *A Canadian company called CarbonCure* Bronte Lord, "This Concrete (Yes, Concrete) Is Going High-Tech," *CNN Business*, July 6, 2018, https://money.cnn.com/2018/06/12/technology/concrete-carboncure/index.html.

169 *Low-carbon-baseload* Michaja Pehl, Anders Arvesen, Florian Humpenoder, Alexander Popp, Edgar Hertwich, and Gunnar Luderer, "Understanding Future Emissions from Low-Carbon Power Systems by Integration of Life-Cycle Assessment and Integrated Energy Modeling," *Nature Energy* 2 (2017): 939–945, https://doi.org/10.1038/s41560-017-0032-9.

169 *Iceland produces* Halldor Armannsson, Thrainn Fridriksson, and Bjarni Reyr Kristjansson, "CO_2 Emissions from Geothermal Power Plants and Natural Geothermal Activity in Iceland," *Geothermics* 34, no. 3 (2005): 286–296, https://doi.org/10.1016/j.geothermics.2004.11.005.

169 *produces almost 2,000 million tons* U.S. Environmental Protection Agency, "Greenhouse Gas Emissions: Inventory of U.S. Greenhouse Gas Emissions and Sinks," https://www.epa.gov/ghgemissions/inventory-us-greenhouse-gas-emissions-and-sinks.

169 *the price of lithium-ion batteries fell* BloombergNEF, "Energy Storage Investments Boom as Battery Costs Halve in the Next Decade," July 31, 2019, https://about.bnef.com/blog/energy-storage-investments-boom-battery-costs-halve-next-decade/.

169 *Los Angeles approved a record-breaking deal* Sammy Roth, "Los Angeles OKs a Deal for Record-Cheap Solar Power and Battery Storage," *Los Angeles Times*, September 10, 2019, https://www.latimes.com/environment/story/2019-09-10/ladwp-votes-on-eland-solar-contract.

169 *the DeGrussa copper and gold mine* Dale Benton, "Australia's Largest Ever Solar Power System Is Up and Running in DeGrussa Gold-Copper Mine," *Mining*, June 8, 2020, https://www.miningglobal.com/supply-chain-and-operations/australias-largest-ever-solar-power-system-and-running-degrussa-gold-copper-mine.

169 *as one headline put it* Akela Lacy, "South Carolina spent $9 billion to dig a hole in the ground and then fill it back in," *The Intercept*, February 6, 2019, https://theintercept.com/2019/02/06/south-caroline-green-new-deal-south-carolina-nuclear-energy/.

170 *distributed and smart grids* National Renewable Energy Laboratory, "Distributed Optimization and Control," https://www.nrel.gov/grid/distributed-optimization-control.html, accessed November 2020.

170 *the largest ground-source closed system* Ball State University, "Geothermal Energy System," https://www.bsu.edu/About/Geothermal.

170 *emissions in Juneau, Alaska* https://juneaucarbonoffset.org/.

170 *New developments in modular "micro-nuclear"* https://www.nuscale power.com/, accessed September 2020.

170 *a modular nuclear plant at Idaho National Laboratory* James Conca, "America Steps Forward to Expand Nuclear Power," *Forbes*, October 21, 2020, https://www.forbes.com/sites/jamesconca/2020/10/21/america -steps-forward-to-expand-nuclear-power/?fbclid=IwAR1ueGKTP8H0 Hh352rnunceTK0Sco-ZxtrX9Ur_IVqQor1Lnf82SoE8PnTk&sh=178b5 aeb72be.

170 *Rolls-Royce is planning to build fifteen mini-reactors* https://www.rolls -royce.com/products-and-services/nuclear/small-modular-reactors .aspx#/, accessed September 2020.

170 *the International Thermonuclear Experimental Reactor* Daniel Kramer, "ITER Disputes DOE's Cost Estimate of Fusion Project, *Physics Today*, April 2018, https://physicstoday.scitation.org/do/10.1063/PT.6.2 .20180416a/full/.

171 *China turned on its own new experimental fusion reactor* "China Turns On Nuclear-Powered 'Artificial Sun,'" Phys.org, December 4, 2020, https:// phys.org/news/2020-12-china-nuclear-powered-artificial-sun.html.

CHAPTER 16

173 *he recounts how* Michael Webber, *Power Trip: The Story of Energy* (New York: Basic Books/Perseus Books, 2019).

174 *could cut emissions* https://www.regi.com/cleaner-fuels/basics.

174 *shipping makes up about 2 to 3 percent* International Marine Organization, Third GHG Study, http://www.imo.org/en/OurWork/Environment /PollutionPrevention/AirPollution/Documents/Third%20Greenhouse %20Gas%20Study/GHG3%20Executive%20Summary%20and%20Re port.pdf.

175 *Aviation is responsible* David Lee, David Fahey, Agnieszka Skowron, Myles Allen, Ulrike Burkhardt, Q. Chen, Sarah Doherty, et al., "The contribution of global aviation to anthropogenic climate forcing from 2000 to 2018," *Atmospheric Environment* 244 (2021): 117834, https://doi .org/10.1016/j.atmosenv.2020.117834.

175 *reducing shipping emissions at least 50 percent* European Commission, "Reducing Emissions from the Shipping Sector," https://ec.europa.eu /clima/policies/transport/shipping_en.

175 *the first one set sail in 2019* Victoria Klesty, "First Battery-Powered Cruise Ship Sails for the Arctic," Reuters, July 1, 2019, https://www.reuters.com /article/us-shipping-electric/first-battery-powered-cruise-ship-sails-for -the-arctic-idUSKCN1TW27E.

175 *retrofit existing cargo ships with masts* Jeff Spross, "Why Cargo Ships Might (Literally) Sail the High Seas Again, "*The Week*, February 26, 2019," https://theweek.com/articles/825647/why-cargo-ships-might-literally -sail-high-seas-again.

175 *only one-third of its warming effect on climate* David Lee, David Fahey, Agnieszka Skowron, Myles Allen, Ulrike Burkhardt, Q. Chen, Sarah Doherty, et al., "The Contribution of Global Aviation to Anthropogenic Climate Forcing for 2000 to 2018," *Atmospheric Environment* 244 (2021):117834, https://doi.org/10.1016/j.atmosenv.2020.117834.

175 *Eviation has created a prototype* Tara Patel, "Electric Planes to Debut for Airline Serving Nantucket, Vineyard," *Bloomberg*, June 18, 2019, https:// www.industryweek.com/technology-and-iiot/article/22027767/electric -planes-to-debut-for-airline-serving-nantucket-vineyard.

175 *known as Wright Electric* Wright Electric, https://weflywright.com/, accessed September 2020.

175 *engineer Duncan Walker calculated* Duncan Walker, "Electric Planes Are Here—But They Won't Solve Flying's CO_2 Problem," *The Conversation*, November 5, 2019, https://theconversation.com/electric-planes -are-here-but-they-wont-solve-flyings-co-problem-125900.

176 *chemical engineers from University College London* "UCL Academics Win British Airways' Sustainable Aviation Fuels Academic Challenge," University College London, May 7, 2019, https://www.ucl.ac.uk/news/2019 /may/ucl-academics-win-british-airways-sustainable-aviation-fuels-aca demic-challenge.

176 *France and the Netherlands required Air France and KLM* David Meyer, "Airline Bailouts Highlight the Debate Over How Green the Coronavirus Recovery Should Be," *Fortune*, June 27, 2020, https://fortune .com/2020/06/27/airline-bailouts-green-pandemic-recovery/.

176 *United Airlines has been refueling* "Expanding Our Commitment to Powering More Flights with Biofuel," United Airlines, May 22, 2019, https:// hub.united.com/united-biofuel-commitment-world-energy-2635867299 .html.

176 *"We [in the aviation industry]"* Russell Hotten, "Dubai Air Show: Emirates Boss Says He Took Too Long to Accept Climate Crisis," *BBC News*, November 20, 2019, https://www.bbc.com/news/business-50481107.

177 *"Farmers and rural Americans"* Justin Worland, "How Climate Change in Iowa is Changing U.S. Politics," *TIME*, Sep 12, 2019, https://time .com/5669023/iowa-farmers-climate-policy/.

177 *24 percent of the global total* U.S. Environmental Protection Agency, "Global Greenhouse Gas Emissions Data," https://www.epa.gov/ghgemis sions/global-greenhouse-gas-emissions-data.

178 *between $2 and $3 trillion* Project Drawdown, "Conservation Agriculture," https://www.drawdown.org/solutions/conservation-agriculture.

178 *In 2020 they won their category* Jayme Lozano, "Lubbock Area Middle School Students Win National Awards for STEM Competition," *Lubbock Avalanche Journal*, July 5, 2020, https://www.lubbockonline.com /news/20200705/lubbock-area-middle-school-students-win-national -awards-for-stem-competition.

179 *a new type of biologically active compost* https://symsoil.com/category /soil-food-web/, accessed September 2020.

180 *rocks that then react with CO_2* Corey Myers and Takao Nakagaki, "Direct Mineralization of Atmospheric CO_2 Using Natural Rocks in Japan," *Environmental Research Letters* 15, no. 12 (2020): https://doi .org/10.1088/1748-9326/abc217.

180 *by adding an enzyme* Adam Subhas, Jess Adkins, Sijia Dong, Nick Rollins, and William Berelson, "The Carbonic Anhydrase Activity of Sinking and Suspended Particles in the North Pacific Ocean," *Limnology and Oceanography* 65, no. 3 (2019): 637–651, https://doi.org/10.1002/lno.11332.

180 *researchers in synthetic biology* Alex Orlando, "Scientists Just Created a Bacteria That Eats CO_2 to Reduce Greenhouse Gases," *Discover*, November 27, 2019, https://www.discovermagazine.com/environment/scien tists-just-created-a-bacteria-that-eats-co2-to-reduce-greenhouse-gases.

180 *Climeworks is a small Swiss company* https://www.climeworks.com/, accessed September 2020.

180 *they've turned to products* Maria Gallucci, "Capture Carbon in Concrete Made with CO_2," *IEEE Spectrum*, February 7, 2020, https://spectrum.ieee .org/energywise/energy/fossil-fuels/carbon-capture-power-plant-co2 -concrete.

181 *to the oil and gas industry* Jesse Jenkins, *Financing mega-scale energy projects: A case study of the Petra Nova carbon capture program*, Paulson Institute, October 2015, https://energy.mit.edu/news/a-case-study-of -the-petra-nova-carbon-capture-project/.

181 *always a more expensive option* June Sekera and Andreas Lichtenberger, "Assessing Carbon Capture: Public Policy, Science, and Societal Need," *Biophysical Economics and Resource Quality* 5, no. 3 (2020): 1–28, https:// doi.org/10.1007/s41247-020-00080-5.

181 *liquid fuel. When burned, it can be carbon neutral* https://carbonengineer ing.com/, accessed September 2020.

181 *2019 study claimed that planting a trillion trees* Thomas Crowther, *Understanding Carbon Cycle Feedbacks to Predict Climate Change at Large Scale*, AAAS Annual Meeting, February 16, 2019, https://www.eurekalert .org/pub_releases/2019-02/ez-pcc021119.php.

181 *the team has already planted over 22 million trees* https://teamtrees.org/, accessed September 2020.

181 *Trillion Trees initiative* http://1t.org.

182 *Ant Forest* Ant Group, "Alipay Gallery: Ant Forest Tree-Planting Spring 2019," *Medium*, April 30, 2019, https://medium.com/alipay-and-the-world /alipay-gallery-ant-forest-tree-planting-spring-2019-dc4e0578cc7c.

182 *Correcting them shows* Joseph Veldman, Julie Aleman, Swanni Alvarado, Michael Anderson, Sally Archibald et al., "Comment on 'The global tree restoration potential,'" Science, 366, no. 6463 (2019), https://doi .org/0.1126/science.aay7976.

182 *Africa Forest Carbon Catalyst* The Nature Conservancy, https://www.na ture.org/en-us/about-us/where-we-work/africa/forest-carbon-catalyst/.

182 *sixty-three cities have signed up* https://cities4forests.com/, accessed September 2020.

182 *over a third of the reductions* Bronson Griscom, Justin Adams, Peter Ellis, Richard Houghton, Guy Lomax, Daniela Miteva, William Schlesinger et al., "Natural climate solutions," *Proceedings of the National Academy of Sciences* 114, no.44 (2017): 11645-11650, https://doi.org/10.1073/pnas .1710465114.

183 *there is no parallel for the sheer volume* Katharine Hayhoe and Robert Kopp, "What Surprises Lurk Within the Climate System?," *Environmental Research Letters* 11, no. 12 (2016): https://doi.org/10.1088/1748-9326.

184 *30 percent more acidic* Richard Feely, Christopher Sabine, Kitack Lee, Will Berelson, Joanie Kleypas, Victoria Fabry and Frank Millero,"Impact of anthropogenic CO_2 on the $CaCO_3$ system in the oceans," *Science* 305, no. 5682 (2004):362–366, https://doi.org/10.1126/science.1097329.

CHAPTER 17

185 *"The planet will survive"* Christiana Figueres and Tom Rivett-Carnac, *The Future We Choose: Surviving the Climate Crisis* (New York: Alfred A. Knopf, 2020).

185 *We don't have all the technology* Solomon Goldstein-Rose, *The 100% Solution: A Plan for Solving Climate Change* (New York: Melville House, 2020).

186 *ten times slower than what's needed* James Temple, "At This Rate, It's Going to Take Nearly 400 Years to Transform the Energy System," *MIT Technology Review*, March 14, 2018, https://www.technologyreview .com/2018/03/14/67154/at-this-rate-its-going-to-take-nearly-400-years -to-transform-the-energy-system/.

186 *American Medical Association* American Association of Public Health Physicians, "American Medical Association Resolution: Divest from Fos-

sil Fuels," June 21, 2018, https://medsocietiesforclimatehealth.org/medi
cal-society-policy-statements/ama-resolution-divest-fossil-fuels/.

186 *British Medical Association* Brian Owens,. 2014, "BMA votes to end investment in fossil fuels," CMAJ. 186(12): E442, https://doi.org/10.1503
/cmaj.109-4857.

187 *it would no longer invest in Arctic oil* Goldman Sachs, "Environmental Policy Framework: D. Climate Change Guidelines," December 2019, https://
www.goldmansachs.com/s/environmental-policy-framework/#climateCh
angeGuidelines.

187 *Michael Corbat, the CEO of Citigroup* Jennifer Surane, "Citi CEO Says
Banks Must Walk If Clients Won't Reduce Emissions," *Bloomberg Green*,
August 19, 2020, https://www.bloomberg.com/news/articles/2020-08-19
/citi-ceo-says-banks-must-walk-if-clients-won-t-reduce-emissions.

187 *twelve more major cities around the world* C40 Cities, "Divesting from
Fossil Fuels, Investing in a Sustainable Future Declaration," September 22,
2020, https://www.c40.org/press_releases/cities-commit-divest-invest.

188 *As of 2020* https://gofossilfree.org/divestment/commitments/, accessed
February 2021.

188 *a significant proportion* Michael Jakob and Jerome Hilaire, "Climate Science: Unburnable Fossil-Fuel Reserves," *Nature* 517 (2015): 150–152,
https://doi.org/10.1038/517150a.

188 *up to $4 trillion* "Stranded Assets and Renewables: How the Energy Tran-
sition Affects the Value of Energy Reserves, Buildings and Capital Stock,"
International Renewable Energy Agency (IRENA), 2017, www.irena.org
/remap.

188 *"climate change poses a major risk"* Climate-Related Market Risk Sub-
committee, Market Risk Advisory Committee of the U.S. Commodity Futures Trading Commission, *Managing Climate Risk in the U.S.
Financial System*, September 2020, https://www.cftc.gov/sites/default
/files/2020-09/9-9-20%20Report%20of%20the%20Subcommittee%20
on%20Climate-Related%20Market%20Risk%20-%20Managing%20
Climate%20Risk%20in%20the%20U.S.%20Financial%20System%20
for%20posting.pdf.

189 *"understand the exposure of your operations"* VeRisk Maplecroft, "Climate Change Vulnerability Index," https://www.maplecroft.com/risk-in
dices/climate-change-vulnerability-index/, accessed September 2020.

189 *a January 2020 letter to global CEOs* Larry Fink, "A Fundamental Reshap-
ing of Finance," BlackRock, January 2020, https://www.blackrock.com/us
/individual/larry-fink-ceo-letter.

189 *Chilean engineer Luis Cifuentes* Luis Cifuentes, Victor Borja-Aburto, Nel-
son Gouveia, George Thurson, and Devra Lee Davis, "Climate Change:

Hidden Health Benefits of Greenhouse Gas Mitigation," *Science* 293, no. 5533 (2001): 1257–1259, https://doi.org/10.1126/science.1063357.

189 *Morgan Stanley estimated* Andrew Harmstone, "Five Sectors That Cannot Escape Climate Change," Morgan Stanley, March 23, 2020, https://www.morganstanley.com/im/en-us/individual-investor/insights/articles/five-sectors-that-cannot-escape-climate-change.html.

189 *estimate by Australian academics* Tom Kompas, Van Ha Pham, and Tuong Nhu Che, "The Effects of Climate Change on GDP by Country and the Global Economic Gains from Complying with the Paris Climate Accord," *Earth's Future*, 2018, https://doi.org/10.1029/2018EF000922.

190 *national GDP would drop by 10 percent* Solomon Hsiang, Robert Kopp, Amir Jina, James Rising, Michael Delgado, Shashank Mohan, D. J. Rasmussen et al., "Estimating Economic Damage from Climate Change in the United States," *Science* 356, no. 6345 (2017): 1362–1369, https://doi.org/10.1126/science.aal4369.

190 *countries like Canada, Japan, and New Zealand* Matthew Kahn, Kamiar Mohaddes, Ryan N. C. Ng, M. Hashem Pesaran, Mehdi Raissi, and Jui-Chung Yang, "Long-Term Macroeconomic Effects of Climate Change: A Cross-Country Analysis," NBER working paper, 2019, doi: 10.3386/w26167.

190 *$22 trillion* Gita Gopinath, "A Race Between Vaccines and the Virus as Recoveries Diverge," International Monetary Fund, January 26, 2021, https://blogs.imf.org/2021/01/26/a-race-between-vaccines-and-the-virus-as-recoveries-diverge/.

190 *"In the 15 countries that emit"* World Health Organization, *COP24 Special Report: Health and Climate Change*, December 5, 2018, https://unfccc.int/sites/default/files/resource/WHO%20COP24%20Special%20Report_final.pdf.

190 *$1.6 to $3.8 trillion per year* United Nations Environment Program, *Emissions Gap Report*, November 26, 2019, https://www.unenvironment.org/resources/emissions-gap-report-2019.

191 *"John D. Rockefeller, the founder of Standard Oil"* Reuters, "Philanthropies, Including Rockefellers, and Investors Pledge $50 Billion Fossil Fuel Divestment," *Scientific American*, September 22, 2014, https://www.scientificamerican.com/article/philanthropies-including-rockefellers-and-investors-pledge-50-billion-fossil-fuel-divestment1/.

SECTION 5:
YOU CAN MAKE A DIFFERENCE.

CHAPTER 18

197 *"more like outbreaks of measles"* Robert Frank, *Under the Influence: Putting Peer Pressure to Work* (Cambridge, MA: MIT Press, 2020).

197 *two geographers noticed solar panels* Marcello Graziano and Kenneth Gillingham, "Spatial Patterns of Solar Photovoltaic System Adoption: The Influence of Neighbors and the Built Environment," *Journal of Economic Geography* 15, no. 4 (2014): 815–839, https://doi.org/10.1093/jeg/lbu036.

197 *Early adopters of solar panels* Kimberly Wolske, Paul Stern, and Thomas Dietz, "Explaining Interest in Adopting Residential Solar Photovoltaic Systems in the United States: Towards an Integration of Behavioral Theories," *Energy Research & Social Science* 25 (2017): 134–151, https://doi .org/10.1016/j.erss.2016.12.023.

197 *ready and eager to bend your ear* Varun Rai and Scott Robinson, "Agent-Based Modeling of Energy Technology Adoption: Empirical Integration of Social, Behavioral, Economic, and Environmental Factors," *Environmental Modelling & Software* 70 (2015): 163–177, https://doi.org/10.1016/j .envsoft.2015.04.014.

197 *in Sweden* Alvar Palm, "Peer Effects in Residential Solar Photovoltaics Adoption: A Mixed Methods Study of Swedish Users," *Energy Research & Social Science* 26 (2017): 1–10, https://doi.org/10.1016/j.erss.2017.01.008.

197 *in China* Teng Zhao, Ziqiang Zhou, Yan Zhang, Ping Ling, and Yingjie Tian, "Spatio-Temporal Analysis and Forecasting of Distributed PV Systems Diffusion: A Case Study of Shanghai Using a Data-Driven Approach," *IEEE Xplore* 5 (2017): 5135–5148, https://doi.org/10.1109/AC CESS.2017.2694009.

197 *in Germany* Johannes Rode and Alexander Weber,. "Does Localized Imitation Drive Technology Adoption? A Case Study on Rooftop Photovoltaic Systems in Germany," *Journal of Environmental Economics and Management* 78 (2016): 38–48, https://doi.org/10.1016/j.jeem.2016.02.001.

198 *Swiss study* Hans Christoph Curtius, Stefanie Lena Hille, Christian Berger, Ulf Joachim Jonas Hahnel, and Rolf Wustenhagen, "Shotgun or Snowball Approach? Accelerating the Diffusion of Rooftop Solar Photovoltaics Through Peer Effects and Social Norms," *Energy Policy* 118 (2018): 596–602, https://doi.org/10.1016/j.enpol.2018.04.005.

198 *21 percent of homes in Australia* Australian Renewable Energy Agency, https://arena.gov.au/renewable-energy/solar/, accessed September 2020.

198 *the city of South Miami* Bobby Magill, "Miami Makes Solar Mandatory for

New Houses," GreenBiz, Jul 26, 2017, https://www.greenbiz.com/article/miami-makes-solar-mandatory-new-houses.

198 *the global rooftop solar industry* Global Industry Analysts, *Global Rooftop Solar PV Industry*, April 2021, https://www.reportlinker.com/p05959926/Global-Rooftop-Solar-PV-Industry.html?utm_source=GNW.

198 *area currently being used* Dave Merrill and Lauren Leatherby, "Here's How America Uses Its Land," *Bloomberg*, July 31, 2018, https://www.bloomberg.com/graphics/2018-us-land-use/.

198 *Invenergy project near Dallas* Darrell Proctor, "Invenergy Unveils Plan for Largest U.S. Solar Project," Power News & Technology for the Global Energy Industry, November 25, 2020, https://www.powermag.com/invenergy-unveils-plan-for-largest-u-s-solar-project/.

198 *second largest solar producing state* Solar Energy Industries Association, "Solar State By State," 2020, https://seia.org/states-map.

198 *In Texas we didn't have enough* Varun Rai and Ariane Beck, "Public Perceptions and Information Gaps in Solar Energy in Texas," *Environmental Research Letters* 10, no. 7 (2015), https://doi.org/10.1088/1748-9326.

199 *this company took in oil workers* "Unemployed Oil Workers Find New Home in Solar Industry," *MarketPlace*, June 7, 2016, https://www.marketplace.org/2016/06/07/unemployed-oil-workers-find-new-home-solar-industry/.

200 *"the belief in one's capabilities"* Albert Bandura, "Self-Efficacy: Toward a Unifying Theory of Behavioral Change," *Psychological Review* 84, no. 2 (1977): 191–215, https://doi.org/10.1037/0033-295X.84.2.191.

200 *people's sense of efficacy YouGov International Climate Change Survey*, June–July 2019, https://d25d2506sfb94s.cloudfront.net/cumulus_uploads/document/7m7cjxikzo/YouGov%20-%20International%20climate%20change%20survey.pdf.

200 *50 percent of Americans feel helpless* A. Leiserowitz, E. Maibach, S. Rosenthal, J. Kotcher, P. Bergquist, M. Ballew, M. Goldberg, and A. Gustafson, "Climate Change in the American Mind," Yale Program on Climate Change Communication, November 2019, https://climatecommunication.yale.edu/wp-content/uploads/2019/12/Climate_Change_American_Mind_November_2019b.pdf.

200 *"don't know where to start"* American Psychological Association, "Majority of US Adults Believe Climate Change Is Most Important Issue Today," February 6, 2020, https://www.apa.org/news/press/releases/2020/02/climate-change

200 *your average-sized dog* Pim Martens, Bingtao Su, and Samantha Deblomme, "The Ecological Paw Print of Companion Dogs and Cats," *BioScience*. 69, no. 6 (2019): 467–474, https://doi.org/10.1093/biosci/biz044.

201 *clinical psychologist Rubin Khoddam points out* Rubin Khoddam, "The Myth of Motivation," *Psychology Today*, August 2017, https://www.psy chologytoday.com/us/blog/the-addiction-connection/201708/the-myth -motivation.

202 *makes us not only more likely to act* Subaru Ken Muroi and Edoardo Bertone, "From Thoughts to Actions: The Importance of Climate Change Education in Enhancing Students' Self-Efficacy," *Australian Journal of Environmental Education* (2019), doi.org/10.1017/aee.2019.12.

202 *to support others who do* Philipp Jugert, Katharine Greenaway, Markus Barth, Ronja Buchner, Sarah Eisentraut, and Immo Fritsche, "Collective Efficacy Increases Pro-Environmental Intentions Through Increasing Self-Efficacy," *Journal of Environmental Psychology* 48 (2016): 12–23, https://doi.org/10.1016/j.jenvp.2016.08.003.

202 *young people who are anxious* Marvin Helferich, Rouven Doran, Daniel Hanss, and Jana Kohler, "Associations Between Climate Change–Related Efficacy Beliefs, Social Norms, and Climate Anxiety Among Young People in Germany," presented at EARA (European Association for Research on Adolescence) Conference, 2020.

202 *affected by the Deepwater Horizon oil spill* John Kaufman, Zachary Goldman, Danielle Sharpe, Amy Wolkin, and Matthew Gribble, "Mechanisms of Resiliency Against Depression Following the Deepwater Horizon Oil Spill," *Journal of Environmental Psychology* 65 (2019): 101329, https://doi.org/10.1016/j.jenvp.2019.101329.

202 *Lisa Altieri's BrightAction* https://brightaction.app/.

205 *connecting with others imbues us* Kathryn Doherty and Thomas Webler, "Social Norms and Efficacy Beliefs Drive the Alarmed Segment's Public-Sphere Climate Actions," *Nature Climate Change* 6 (2016): 879–884, https://doi.org/10.1038/nclimate3025.

CHAPTER 19

207 *"Eating organic is nice"* David Wallace-Wells, *The Uninhabitable Earth: Life After Warming* (New York: Tim Duggan Books, 2019).

207 *climate scientists who take their own carbon footprint* Shahzeen Attari, David Krantz, and Elke Weber, "Climate Change Communicators' Carbon Footprints Affect Their Audience's Policy Support," *Climatic Change* 154 (2019): 529–545, https://doi.org/10.1007/s10584-019-02463-0.

208 *washing your clothes in cold water* Pierre Delforge, "Home Idle Load: Devices Wasting Huge Amounts of Electricity When Not in Active Use," National Resources Defense Council Issue Paper, July 14, 2015, https://www

.nrdc.org/resources/home-idle-load-devices-wasting-huge-amounts
-electricity-when-not-active-use.

208 *some $19 billion* Pierre Delforge, "Home Idle Load: Devices Wasting Huge
Amounts of Electricity When Not In Active Use," *National Resources Defense Council,* July 14, 2015, https://www.nrdc.org/resources/home-idle
-load-devices-wasting-huge-amounts-electricity-when-not-active-use.

209 *a third of the food grown* Project Drawdown, "Food Waste," https://
www.drawdown.org/solutions/reduced-food-waste, accessed September 2020.

209 *people throw away enough food* Jimmy Nguyen, USDA Food Waste Challenge Team, "Creative Solutions to Ending School Food Waste," USDA,
August 26, 2014, https://www.usda.gov/media/blog/2014/08/26/creative
-solutions-ending-school-food-waste.

209 *Second Harvest estimates 58 percent* Second Harvest, "The Avoidable
Crisis of Food Waste," https://secondharvest.ca/research/the-avoidable
-crisis-of-food-waste/.

209 *"It'll actually dictate what I buy for dinner"* "Food Waste: There's an App
for That (a Few, Actually)," *CBC News,* October 10, 2019, https://www
.cbc.ca/news/technology/what-on-earth-newsletter-thanksgiving-food
-waste-app-1.5316720.

210 *Eating lower down in the food chain* Joseph Poore and Anthony Nemecek,
"Reducing Food's Environmental Impacts Through Producers and Consumers," *Science* 360, no. 6392 (2018): 987–992, https://doi.org/10.1126
/science.aaq0216.

210 *more than 30 billion* Food and Agriculture Organization of the United
Nations, "FAOSTAT: Live Animals," (2019) http://www.fao.org/faostat
/en/#data/QA.

210 *Account for 14 percent* United Nations Food and Agriculture Organization, *Tackling climate change through livestock,* October 21, 2014, http://
www.fao.org/ag/againfo/resources/en/publications/tackling_climate
_change/index.htm.

210 *Per kilogram* Joseph Poore and Anthony Nemecek, "Reducing Food's Environmental Impacts Through Producers and Consumers," Science 360,
no. 6392 (2018): 987–992, https://doi.org/10.1126/science.aaq0216.

210 *Broken Arrow Ranch* https://brokenarrowranch.com/.

211 *A Futurra start-up* https://www.lovebugpetfood.com/.

211 *Purina called RootLab* Corinne Gretler and Deena Shankar, "Purina
Wants to Feed Your Dog Crickets and Fish Heads," *Bloomberg,* January 29, 2019, https://www.bloomberg.com/news/articles/2019-01-29
/purina-wants-your-dog-to-save-the-planet-by-eating-fish-heads.

211 *household garbage increased under coronavirus* Yelena Dzhanova, "Sanitation Workers Battle Higher Waste Levels in Residential Areas as Coronavirus Outbreak Persists," CNBC, May 16, 2020, https://www.cnbc.com/2020/05/16 /coronavirus-sanitation-workers-battle-higher-waste-levels.html.

211 *supply chains for recycling were disrupted* Scott Horsley, "'Hard, Dirty Job': Cities Struggle to Clear Garbage Glut in Stay-at-Home World," *Morning Edition*, NPR, September 21, 2020, https://www.npr .org/2020/09/21/914029452/hard-dirty-job-cities-struggle-to-clear-gar bage-glut-in-stay-at-home-world.

212 *the highest-impact personal carbon emission reduction* Kim Nicholas, "Personal Choices to Reduce Your Contribution to Climate Change," http://www.kimnicholas.com/uploads/2/5/7/6/25766487/fig1full.jpg.

212 *Malcolm Gladwell says* Malcolm Gladwell, *The Tipping Point: How Little Things Can Make a Big Difference* (New York: Little, Brown and Company, 2000).

212 *Michael Mann wrote in his 2019* TIME *essay* Michael Mann, "Lifestyle Changes Aren't Enough to Save the Planet. Here's What Could," *TIME*, September 12, 2019, https://time.com/5669071/lifestyle-changes-climate -change/.

CHAPTER 20

215 *"If norms lead people"* Cass Sunstein, *How Change Happens* (Cambridge: MIT Press, 2019).

216 *new environment and sustainability strategy* Wandsworth Borough Council, *Wandsworth Environmental and Sustainability Strategy 2019– 2030*, https://wandsworth.gov.uk/media/6769/wandsworth_environment _and_sustainability_strategy_2019_30.pdf.

217 *They want to speak up* Nathaniel Geiger, "Untangling the Components of Hope: Increasing Pathways (Not Agency) Explains the Success of an Intervention That Increases Educators' Climate Change Discussions," *Journal of Environmental Psychology* 66 (2019): 101366, https://doi.org/10.1016/j .jenvp.2019.101366.

217 *According to polling data* Jennifer Marlon, Peter Howe, Matto Mildenberger, Anthony Leiserowitz, and Xinran Wang, "Yale Climate Opinion Maps 2020," September 2, 2020, https://climatecommunication.yale.edu /visualizations-data/ycom-us/.

217 *Ed Maibach has been saying* Edward Maibach, "Increasing Public Awareness and Facilitating Behavior Change: Two Guiding Heuristics," in *Biodiversity and Climate Change: Transforming the Biosphere* eds. Thomas Lovejoy and Lee Hannah (New Haven: Yale University Press, 2019).

217 *your brain waves start to synchronize* Elena Renken, "How Stories Connect and Persuade Us: Unleashing the Brain Power of Narrative," 89.3 KPPC, April 11, 2020, https://www.scpr.org/news/2020/04/11/91857/how-stories-connect-and-persuade-us-unleashing-the/.

218 *"any company that hasn't already got a net-zero"* Robin Pomeroy, "COVID, Climate and Inequality: This Week's Great Reset Podcast," World Economic Forum, September 25, 2020, https://www.weforum.org/agenda/2020/09/covid-climate-and-inequality-this-weeks-great-reset-podcast/.

218 *"As awful as the COVID pandemic is"* Bill Gates, "COVID-19 Is Awful. Climate Change Could Be Worse," August 4, 2020, https://www.gatesnotes.com/Energy/Climate-and-COVID-19.

218 *Admiral Samuel Locklear III* Bryan Bender, "Chief of US Pacific forces calls climate change biggest worry," *Boston Globe*, March 9, 2013, https://www.bostonglobe.com/news/nation/2013/03/09/admiral-samuel-locklear-commander-pacific-forces-warns-that-climate-change-top-threat/BHdPVCLrWEMxRe9IXJZcHL/story.html.

218 *"Drought and severe storms are triggering"* "General Ron Keys: Air Force General (ret'd), Climate Advocate," New Climate Voices, http://www.newclimatevoices.org/ron-keys/.

219 *"COVID will eventually end"* Amanda Millstein, "Opinion: It's Climate Change That Keeps This Bay Area Doctor Up at Night," *Mercury News*, September 24, 2020, https://www.mercurynews.com/2020/09/24/opinion-its-climate-change-that-keeps-this-doctor-up-at-night/.

219 *a third of TV weather forecasters* Edward Maibach, Sara Cobb, Erin Peters, Carole Mandryk, David Straus, Dann Sklarew, et al., "A National Survey of Television Meteorologists About Climate Change: Education," George Mason University Center for Climate Change Communication, 2011, http://climatechangecommunication.org/wp-content/uploads/2016/03/June-2011-A-National-Survey-of-Television-Meterologists-about-Climate-Change-Education.pdf.

219 *In media markets where the weather forecasters* Teresa Myers, Edward Maibach, Bernadette Woods Placky, Kimberley Henry, Michael Slater, and Keith Seitter, "Impact of the Climate Matters Program on Public Understanding of Climate Change," *Weather, Climate, and Society* 12, no. 4 (2020): 863–876, https://doi.org/10.1175/WCAS-D-20-0026.1.

219 *the majority of U.S. broadcast meteorologists now agree* David Perkins, Kristin Timm, Teresa Myers, and Edward Maibach, "Broadcast Meteorologists' Views on Climate Change: A State-of-the-Community Review," *Weather, Climate, and Society* 12, no. 2 (2020): 249–262, https://doi.org/10.1175/WCAS-D-19-0003.1.

221 *people who cared about it weren't talking* Meaghan Guckian, *Social Signals for Change: Examining the Role of Interpersonal Communication for Positive Ecological Progress*, University of Massachusetts dissertation, 2019.

222 *"your TED Talk inspired our friend Howard"* Howard Kirkham, "Climate Conversations in the Forest," https://www.oursafetynet.org/2020/03/17 /what-i-learned-when-i-talked-to-strangers-about-the-climate-crisis/.

222 **How to Have Impossible Conversations** Peter Boghossian and James Lindsay, *How to Have Impossible Conversations* (New York: Lifelong Books, 2019).

223 *the simple act of having a conversation* Matthew Goldberg, Sander van der Linden, Edward Maibach, and Anthony Leiserowitz, "Discussing Global Warming Leads to Greater Acceptance of Climate Science," *Proceedings of the National Academy of Sciences* 116, no. 30 (2019): 14804–14805, https:// doi.org/10.1073/pnas.1906589116.

CHAPTER 21

225 *"Most people do not listen"* Stephen Covey, *The 7 Habits of Highly Effective People* (New York: Free Press, 1989).

231 *Project Drawdown* Paul Hawkins, *Drawdown: The Most Comprehensive Plan Ever Proposed to Reverse Global Warming* (New York: Penguin Books, 2017).

232 *Climate Mind* https://climatemind.org/.

235 *"The chief task in life is simply this"* Epictetus, *Discourses*, 2.5.4–5.

236 **keep listening, because the longer you listen** Karin Kirk, "What Conversations with Voters Taught Me About Science Communication," *Scientific American*, November 10, 2020, https://www.scientificamerican.com /article/what-conversations-with-voters-taught-me-about-science-com munication/?amp;text=What.

236 *"For successful dialogue you need to try"* Tania Israel, *Beyond Your Bubble: How to Connect Across the Political Divide* (Washington, D.C.: APA LifeTools, 2020).

236 *Renée Lertzman calls this process "attunement"* Project InsideOut, "Be a Guide: Our Guiding Principles," 2020, https://projectinsideout.net/wp -content/uploads/2020/09/Be-a-Guide.pdf.

236 *"You can't change people's minds"* Jonathan Haidt, *The Righteous Mind: Why Good People Are Divided by Politics and Religion* (New York: Random House, 2012).

237 *"See the experience as a way to learn"* Robin Webster and George Marshall, *Talking Climate Handbook: How to Have Conversations About Climate Change in Your Daily Life* (Oxford: Climate Outreach, 2019), https://

climateoutreach.org/reports/how-to-have-a-climate-change-conversa
tion-talking-climate/.

CHAPTER 22

241 *"It was reasonable to struggle"* P. D. James, *The Children of Men* (New York: Vintage, 2006, reissue).

241 *"When the thought of climate doom arrives"* Jedediah Britton-Purdy, "The Concession to Climate Change I Will Not Make," *Atlantic*, January 6, 2020, https://www.theatlantic.com/science/archive/2020/01/becom ing-parent-age-climate-crisis/604372/.

242 *"No matter how young you are"* Sheena Lyonnais, "Yonge Interviews: 10-Year-Old Activist Hannah Alper," *Yonge Street*, September 11, 2013, http://www.yongestreetmedia.ca/features/hannahalper091113.aspx.

242 *"Whoever sows injustice will reap calamity"* Proverbs 22:8 (New International Version).

242 *"for they have sown the wind"* Hosea 8:7 (King James Version).

242 *we especially underestimate unfamiliar risks* Susana Gouveia and Valerie Clarke, "Optimistic Bias for Negative and Positive Events," *Health Education* 101, no. 5 (2001): 228–234, https://doi.org/10.1108/09654280110402080.

242 *assume we have more control over circumstances* Cynthia Klein and Marie Helweg-Larsen, "Perceived Control and Optimistic Bias: A Meta-Analytic Review," *Psychology and Health* 17, no. 4 (2002): 437–446, https://doi.org /10.1080/0887044022000004920.

242 *"we expect things to turn out better"* Tali Sharot, *The Optimism Bias: A Tour of the Irrationally Positive Brain* (New York: Vintage, 2012).

243 *hope that carry us forward* Simon Bury, Michael Wenzel, and Lydia Woodyatt, "Against the Odds: Hope as an Antecedent of Support for Climate Change Action," *British Journal of Social Psychology* 59, no. 2 (October 7, 2019): 289–310, https://doi.org/10.1111/bjso.12343.

243 *"We're not fighting"* Peter Kalmus, Twitter post, February 2, 2020, 12:13 p.m., https://twitter.com/climatehuman/status/1224018042520137734.

244 *"Active Hope is a practice"* Joanna Macy and Chris Johnstone, *Active Hope: How to Face the Mess We're in Without Going Crazy* (New York: New World Library, 2012).

244 *"We know that troubles help us learn"* Romans 5:3–4 (New Life Version).

INDEX

ABOUT THE AUTHOR

Katharine Hayhoe is a climate scientist and chief scientist for The Nature Conservancy. She is also the Endowed Professor in Public Policy and Public Law and Paul W. Horn Distinguished Professor at Texas Tech University. She has been named a United Nations Champion of the Earth and one of *TIME*'s 100 Most Influential People, and serves as the climate ambassador for the World Evangelical Alliance. Katharine was a lead author for the U.S. Second, Third, and Fourth National Climate Assessments, hosts the PBS digital series *Global Weirding*, and has written for the *New York Times*. Her TED Talk "The Most Important Thing You Can Do to Fight Climate Change: Talk About It" has been viewed over 5 million times. She has a BSc in physics and astronomy from the University of Toronto and an MS and a PhD in atmospheric science from the University of Illinois at Urbana-Champaign.